T0246998

Futures of the Sun

Forerunners: Ideas First

Short books of thought-in-process scholarship, where intense analysis, questioning, and speculation take the lead

FROM THE UNIVERSITY OF MINNESOTA PRESS

Imre Szeman
Futures of the Sun: The Struggle over Renewable Life

Jordan S. Carroll
Speculative Whiteness: Science Fiction and the Alt-Right

Shenila Khoja-Moolji
The Impossibility of Muslim Boyhood

Cait McKinney
I Know You Are, but What Am I? On Pee-wee Herman

Lisa Diedrich
Illness Politics and Hashtag Activism

Mark Foster Gage
On the Appearance of the World: A Future for Aesthetics in Architecture

Tia Trafford
Everything Is Police

EL Putnam
Livestreaming: An Aesthetics and Ethics of Technical Encounter

Dominic Boyer
No More Fossils

Sharad Chari
Gramsci at Sea

Kathryn J. Gindlesparger
Opening Ceremony: Inviting Inclusion into University Governance

J. Logan Smilges
Crip Negativity

Shiloh Krupar
Health Colonialism: Urban Wastelands and Hospital Frontiers

Antero Garcia
All through the Town: The School Bus as Educational Technology

Lydia Pyne
Endlings: Fables for the Anthropocene

Margret Grebowicz
Rescue Me: On Dogs and Their Humans

Sabina Vaught, Bryan McKinley Jones Brayboy, and Jeremiah Chin
The School–Prison Trust

After Oil Collective; Ayesha Vemuri and Darin Barney, Editors
Solarities: Seeking Energy Justice

Arnaud Gerspacher
The Owls Are Not What They Seem: Artist as Ethologist

Tyson E. Lewis and Peter B. Hyland
Studious Drift: Movements and Protocols for a Postdigital Education

Mick Smith and Jason Young
Does the Earth Care? Indifference, Providence, and Provisional Ecology

Caterina Albano
Out of Breath: Vulnerability of Air in Contemporary Art

Gregg Lambert
The World Is Gone: Philosophy in Light of the Pandemic

Grant Farred
Only a Black Athlete Can Save Us Now

Anna Watkins Fisher
Safety Orange

Heather Warren-Crow and Andrea Jonsson
Young-Girls in Echoland: #Theorizing Tiqqun

Joshua Schuster and Derek Woods
Calamity Theory: Three Critiques of Existential Risk

Daniel Bertrand Monk and Andrew Herscher
The Global Shelter Imaginary: IKEA Humanitarianism and Rightless Relief

Catherine Liu
Virtue Hoarders: The Case against the Professional Managerial Class

Christopher Schaberg
Grounded: Perpetual Flight . . . and Then the Pandemic

Marquis Bey
The Problem of the Negro as a Problem for Gender

Cristina Beltrán
Cruelty as Citizenship: How Migrant Suffering Sustains White Democracy

Hil Malatino
Trans Care

Sarah Juliet Lauro
Kill the Overseer! The Gamification of Slave Resistance

(Continued on page 80)

Futures of the Sun
The Struggle over Renewable Life

Imre Szeman

University of Minnesota Press

MINNEAPOLIS

LONDON

A portion of lecture 2 appeared in a different form as "From Steam Fetishism to Solar Fetishism," in *Energy Humanities,* July 14, 2023, energyhumanities.ca.

ISBN 978-1-5179-1769-2 (PB)
ISBN 978-1-4529-7158-2 (Ebook)
ISBN 978-1-4529-7256-5 (Manifold)

Futures of the Sun: The Struggle over Renewable Life by Imre Szeman is licensed under a Creative Commons Attribution-NonCommercial-NoDerivatives 4.0 International License.

Published by the University of Minnesota Press, 2024
111 Third Avenue South, Suite 290
Minneapolis, MN 55401-2520
www.upress.umn.edu

Available as a Manifold edition at manifold.umn.edu

The University of Minnesota is an equal-opportunity educator and employer.

To dms7@

for future energy systems still to come

Let's embrace the idea that there are two commonwealths. The one is vast and truly common to all, and includes the gods as well as mankind; within it, we look neither to this mere corner nor to that, but we measure the boundaries of our state by the sun's course. The other is the one in which we are enrolled by the circumstances of our birth—I mean Athens or Carthage or any other city that belongs not to the whole of mankind but to a particular population. Certain people give devoted service to both commonwealths, the greater and the lesser, at the same time; some serve only the lesser, some only the greater.

—SENECA, *De otio*

Contents

Preface   xiii

Lecture 1. Renewable Futures and the
Temptations of Nationalism   1

Lecture 2. The Life and Times of Bill Gates,
Eco-Warrior   25

Lecture 3. From Convoys to Commons   49

Acknowledgments   79

Preface

WHO OWNS THE SUN? It's a seemingly ridiculous question. One could imagine sci-fi narratives in which moons or planets might eventually become subject to the law of property (subdivisions on the moon! mining on Mars!). But the sun? Impossible! There's just no way one could put up fences on its surface or collect up all the energies it shines out to the universe.

And yet the battle over just who owns the energy of the sun *is* now in the process of being fought out. The practices of energy ownership definitive of the era of fossil fuels are in all too many cases quickly being extended to renewables.[1] The ownership with which I'm concerned in this short book isn't the control of renewable energy technologies, sites, systems, or networks by companies or governments; rather, it's about who is trying to lay claim to the

---

1. See, e.g., Gretchen Bakke, *The Grid: The Fraying Wires between Americans and Our Energy Future* (New York: Bloomsbury, 2017); Dominic Boyer and Cymene Howe, *Wind and Power in the Anthropocene* (Durham, N.C.: Duke University Press, 2019); and David McDermott Hughes, *Who Owns the Wind? Climate Crisis and the Hope of Renewable Energy* (New York: Verso, 2021). See also Daniel M. Berman and John T. O'Connor, *Who Owns the Sun? People, Politics, and the Struggle for a Solar Economy* (White River Junction, Vt.: Chelsea Green, 1996). Berman and O'Connor's overview of the political challenges and opportunities of solar energy remains pertinent today.

narratives guiding our transition from fossil fuels to renewable energy, how they are doing so, and why and to what ends.

The emergence of the sun (and water, wind, and earth) as a source of power that might soon eclipse fossil fuels constitutes a transition whose effects will go well beyond the type of energy we use. Even the International Energy Agency (IEA) recognizes this. The IEA's 2022 World Energy Outlook report projects that owing to the rapid greening of energy systems, CO_2 emissions from the power sector will likely peak in 2025. "Global fossil fuel use has risen alongside gross domestic product (GDP) since the start of the Industrial Revolution in the 18th century," the report notes. "Putting this rise into reverse while continuing to expand the global economy will be a pivotal moment in energy history."[2] For many around the world, this civilizational shift promises new economic, social, and political possibilities; for others, it is a threat to be managed, controlled, or eliminated altogether. Even in the IEA's short summary of where we now find ourselves in energy history, one can find an example of the narratives of transition that interest me and the social and political imaginaries at play within them. For the IEA, the continued expansion of the global economy is a priority above all else, and though the reduction of CO_2 emissions is to be welcomed—how could it not be?—the uncertainties of the post–fossil fuel world are definitely not. These uncertainties go beyond figuring out how to balance economic expansion with climate change mitigation strategies. They extend to what 2025 means for the future of capitalism and to the possibility of deep disruptions to the norms and values that underpinned the production, distribution, and consumption of all things, including energy, from the start of the age of oil up to the present.

The anxiety that the IEA expresses, which positions renewable energies and growth as potential antagonists, is a common one; indeed, it might be seen as the starting point for all narratives intent on explaining the shape our renewable futures will take. The

2. International Energy Agency, *World Outlook Report 2022* (Paris: IEA, 2022), https://www.iea.org/reports/world-energy-outlook-2022.

current forceful pushback on renewables by some governments (the government of the Canadian province of Alberta offers a prominent example) and the fossil fuel industry is due in part to their under-standing of this antagonism as an immutable reality.[3] They argue that growth is impossible without fossil fuels and that working-class jobs depend on the continuation of extraction; for these reasons alone, transition must happen slowly and steadily, if it is to take place at all. But at this point in history, even those governments and fossil fuel companies resistant to energy transition recognize its inevitability, as evidenced by investments in renewables made by both international and national oil companies as early as the 1980s; in the main, delay tactics on transition hope only to push the peak date of 2025 forward in time—to 2030, 2040, or even later—so that in the interim, profit can continue to be reaped on capital invested in extraction.[4] A related narrative proposed by defenders of fossil fuels foregrounds technological innovation as the way to resolve the

3. See Emma Graney, "Alberta to Pause New Solar and Wind Power Projects for Six Months amid Review of End-of-Life Rules," *Globe and Mail,* August 3, 2023, https://www.theglobeandmail.com/business/industry -news/energy-and-resources/article-alberta-to-pause-new-solar-and-wind -power-projects-for-six-months-amid/. Criticism of the decisions made by the new Alberta government on renewables has been almost universal. See Gary Mason, "Alberta's Freeze on Renewable Energy Projects Belongs in the Hall of Fame of Dumb Ideas," *Globe and Mail,* August 9, 2023, https:// www.theglobeandmail.com/opinion/article-albertas-freeze-on-renewable -energy-projects-belongs-in-the-hall-of/.

4. The writing is on the wall for oil. According to the *New York Times,* in 2023, more than US$1.7 billion will be invested in renewables, in com-parison to US$1 billion in fossil fuels. See David Gelles, Brad Plumer, Jim Tankersley, and Jack Ewing, "The Clean Energy Future Is Arriving Faster Than You Think," *New York Times,* August 17, 2023, https://www.nytimes .com/interactive/2023/08/12/climate/clean-energy-us-fossil-fuels.html. In Alberta, major industry players, such as Suncor and Enbridge, have invested significantly in renewables and plan to continue to do so. While some existing oil and gas companies have reaffirmed their ongoing commitment to fossil fuel extraction, many large, international companies, including BP and Shell, have reimagined themselves as energy companies instead of oil and gas companies.

antagonism between GDP growth and CO_2 decline. For those who continue to want to extract, new technologies, such as hydrogen or carbon capture, utilization, and storage (CCUS), offer a means by which the consequences of fossil fuel use can be managed so that old-fashioned growth can continue. Whether CCUS works matters less than the appearance of technology having come to rescue, so as to allow oil to keep flowing smoothly and steadily.

These aren't the only arguments to keep the energy status quo and reap the supposed benefits that accompany it. Other narratives have tried to defuse post-oil uncertainties by showing how the rise of renewables will result in economic growth. Here, too, the promise of innovation plays a key role, with new technologies—ever more efficient solar panels and electric vehicles being the most prominent of these—transforming antagonists into allies. Such narratives position renewables as a way of securing a green, clean future for existing forms of capitalist enterprise. And then there is the opposing position, which sees the threat of climate change and the advent of renewable energies as opportunities to question the legitimacy and necessity of the growth of GDP on into perpetuity. Theorists of degrowth have made powerful arguments against the rationale for measuring human well-being by the metric of economic growth and have proposed ways to envision human flourishing that would not come at the expense of the environment.[5] Some have challenged the capacity of degrowth to manage the trick of improving the quality of life for everyone on the planet, arguing that growth (of hopefully a clean, green kind) remains important to providing greater opportunities for many on the planet who were unable to benefit from the economic outcomes of the fossil fuel era.[6]

5. See Jason Hickel, *Less Is More: How Degrowth Will Save the World* (New York: William Heinemann, 2021), and Matthias Schmelzer, Andrea Vetter, and Aaron Vansintjan, *The Future Is Degrowth: A Guide to a World beyond Capitalism* (New York: Verso, 2022).
6. See Matt Huber, *Climate Change as Class War: Building Socialism on a Warming Planet* (New York: Verso, 2022). Huber has been an outspoken critic of degrowth as a political and economic strategy.

Although I've touched on many of these arguments in this book, my focus is on a far messier set of narratives being written, told, and sold about who should own the sun and what it means to do so. Contemporary forms of political power, whether implicitly or explicitly, have for two centuries grown as dependent on the power of fossil fuels as have economies.[7] As the end of the fossil fuel era approaches, it should thus come as little surprise that competing narratives of energy futures are already in play, each trying to be the first to make sense of what a politics anchored on renewable energy and legitimated by climate action (to whatever degree and extent) might look like. These narratives are not yet fully coherent ones—nor would one expect them to be. They are now in the process of being tested in the space of uncertainty captured in the IEA's assessment of where we find ourselves. In each chapter, I offer an assessment of a still-emerging narrative about the shape of political power after fossil fuels; my own assessment must be seen as similarly emergent—as an attempt to grapple, as much as possible, with phenomena still in the process of development.

It should go without saying: climate change is the most significant crisis faced by modern governments, regardless of whether they take up the challenge of doing anything about it. The growing number of deadly weather-related events has transfigured climate change from an idea articulated by experts into a reality with which it is necessary to contend. Governments might be unnerved by climate change and by renewable energy transition. But they and other actors have asserted or reasserted the legitimacy of their power by making a case for their unique ability to take on the challenge of climate change and energy transition. The post–fossil fuel narra-

7. Dipesh Chakrabarty reminds us that "the mansion of modern freedoms stands on an ever-expanding base of fossil-fuel use. Most of our freedoms so far have been energy intensive." See Chakrabarty, "The Climate of History: Four Theses," *Critical Inquiry* 35, no. 2 (2009): 208. Timothy Mitchell puts it just as directly: "fossil fuels helped create both the possibility of twentieth-century democracy and its limits." Mitchell, "Carbon Democracy," *Economy and Society* 38, no. 2 (2009): 399.

tives I explore here are all attempts to reaffirm status quo forms of political, economic, and technological power. In the first chapter, I look at the way in which advocates for postglobal liberalism (contra populism) argue that liberal nationalism is the only way in which to combat climate change. The second chapter explores the way in which neoliberalism is being extended to decision-making about climate change and energy transition. In this case, the failure of the nation-state to take effective charge of energy transition is being used by private industry and entrepreneurs as further evidence of the need for them to lead the way in shaping what the world will look like post-2025. I end with an assessment of the role energy plays in populist nationalisms. In such nationalisms, energy transition is viewed as a threat to populism's power and legitimacy, even as fossil fuel extraction is foregrounded as an essential element of tradition. I leave it up to the reader to determine whether the brief concluding argument I make for the use of tradition and myth to very different political ends makes sense as a starting point for a new politics of energy transition and climate change. What I know for certain is that there needs to be a counternarrative to the futures of the sun being articulated today.

The degree to which any one of these narratives comes to define the politics of the post–fossil era will depend on its success at becoming the norm against which others are forced to measure themselves. In essence, we can see each as struggling to transform itself into the undisputed "common sense" about how to approach energy transition and climate change. Stuart Hall and Alan O'Shea describe common sense as comprising "frameworks of meaning with which to make sense of the world."[8] Common sense shares similarities with what Antonio Gramsci has described as "spontaneous consent," and perhaps even more so with Louis Althusser's description of ideology as the recognition by subjects "that 'it's

8. Stuart Hall and Alan O'Shea, "Common-sense Neoliberalism," *Soundings: A Journal of Politics and Culture* 55 (2013): 8.

really true,' that 'this is the way it is,' not some other way, that they have to obey God, the priest, De Gaulle, the boss, the engineer, and love their neighbor, and so on . . . they have recognized that 'all is well' (the way it is), and they say, for good measure: *So be it.*"[9] But there is a reason I prefer using common sense as a way to describe positions that subjects and institutions name as irrefutable. The notion of common sense is not just a description of an already achieved state—or not only that—but names the ongoing, active struggle to establish a norm, perhaps especially in situations in which one does not yet exist. Appeals to "common sense" are a way of bringing people into one's narrative fold, an almost effortless way of defining an "us" (those who recognize the given and only way of doing things) versus a "them" (those who, it is implied in the appeal, hold extreme or critical views or who just don't get it).[10] In other words, "common sense" is something that has to be made. Each of the narratives of renewable futures at which I look here makes the case that its views are "common sense" precisely because they are anything but.

The chapters included in this book were presented at the University of Glasgow on November 15, 16, and 17 as the 2022

9. Antonio Gramsci, *Selections from the Prison Notebooks,* ed. Quentin Hoare and Geoffrey Nowell Smith (London: Lawrence and Wishart, 1976), 12; Louis Althusser, *On the Reproduction of Capitalism,* trans. G. M. Goshgarian (New York: Verso, 2014), 197.

10. This is precisely the sense in which Hall and O'Shea use it: "When politicians try to win consent or mobilise support for their policies, they frequently assert that these are endorsed by 'hard-working families up and down the country.' Their policies cannot be impractical, unreasonable or extreme, they imply, because they are solidly in the groove of popular thinking—'what everybody knows,' takes-for-granted and agrees with—the folk wisdom of the age. This claim by the politicians, if correct, confers on their policies popular legitimacy. . . . In fact, what they are really doing is not just invoking popular opinion but *shaping and influencing* it so they can harness it in their favour. By asserting that popular opinion *already agrees,* they hope to produce agreement *as an effect.* This is the circular strategy of the self-fulfilling prophecy." Hall and O'Shea, "Common-sense Neoliberalism," 8.

Leverhulme Lectures. The lectures were written for a mixed audience comprising both academics and nonacademics and consciously written in a style that—I hoped—would appeal to both. Although I would have liked to extend parts of the argument beyond what I could offer in three one-hour lectures, I have chosen to leave the text as close to its original form as possible, in large part to avoid interrupting the flow of a text written to be read aloud. There have been some changes made to address grammatical and spelling mistakes, and small nudges added to keep things flowing, but no other major changes.

Lecture 1. Renewable Futures and the Temptations of Nationalism

Energy Transition and Common Sense

The year 2021 marked the end of the age of oil. Bill McKibben was the first to make this announcement, in a January 2021 *New Yorker* article in which he reviewed the first steps on climate change taken by then new U.S. president Joe Biden.[1] It's a claim that others have made since, including the Toronto *Globe and Mail*'s editorial board, which in May 2021 pondered what the end of the age of oil meant for Canada, a country that has tied its fate to resource extraction.[2] The *Globe* article was written in response to a two-hundred-page report issued earlier that week by the IEA, titled *Net Zero by 2050: A Roadmap for the Global Energy Sector.*[3] The IEA was established

1. Bill McKibben, "The Biden Administration's Landmark Day in the Fight for the Climate," *New Yorker,* January 28, 2021, https://www .newyorker.com/news/daily-comment/the-biden-administrations -landmark-day-in-the-fight-for-the-climate.

2. Editorial Board, "The Age of Oil Is Coming to an End. What Does That Mean for Canada?," *Globe and Mail,* May 22, 2021, https://www .theglobeandmail.com/opinion/editorials/article-the-age-of-oil-is-coming -to-an-end-what-does-that-mean-for-canada/.

3. International Energy Agency, *Net Zero by 2050: A Roadmap for the Global Energy Sector* (Paris: IEA, 2021), https://iea.blob.core.windows .net/assets/deebef5d-0c34-4539-9d0c-10b13d840027/NetZeroby2050 -ARoadmapfortheGlobalEnergySector_CORR.pdf.

by the Organisation for Economic Co-operation and Development in the wake of the 1973 oil crisis and was tasked with providing data to member countries about the state of oil markets. Though the IEA had already moved in recent years to attend to mechanisms for effective energy transition, this new report made it clear that it, too, had given up on oil. In its 2020 World Energy Outlook report, "net zero emissions by 2050" was just one of a series of scenarios examined in the wake of the Covid-19 pandemic. By 2021, a move to renewables had become the only scenario the IEA thought worth considering. It was a shift whose significance was not lost on all those involved in the game of energy. Renewables had won the day; the new story of energy would place them at the center of the action.[4]

This is not to say that fossil fuels will now simply disappear. Those with stakes in fossil fuels and whose economic and political power depends on them do not intend to leave the center stage of history quietly. Fossil fuel countries and companies are already fighting hard on multiple fronts to impede a shift in energy regimes.[5] Even if everyone involved in the fossil fuel industry now knows that the jig is up, they want to make certain to extract all they can as soon as they can—too much capital has been invested in existing oil infrastructure not to do so. Private shareholders want to make sure to get their money back, and petro-governments want to ensure that their capital investments have not gone to waste and that national economies backed by fossil fuel extraction won't collapse.

    4. See Tim Quinson and Mathieu Benhamou, "Five Takeaways from Global Banks' Green vs. Fossil Financing," *Bloomberg Green,* May 18, 2021, https://www.bloomberg.com/news/articles/2021-05-19/jpmorgan-tops-banks-supporting-fossil-fuel-and-signals-green-shift, and Emma Graney and Jeffrey Jones, "Big Oil Loses Carbon Emissions Showdown in Landmark Case," *Globe and Mail,* May 26, 2021, https://www.theglobeandmail.com/business/article-canadas-oil-industry-on-watch-after-dutch-court-orders-shell-to-cut/.
    5. For recent examples, see Dharna Noor, "Mike Rowe's New Discovery+ Show Is Big Oil-Funded Propaganda," *Gizmondo,* April 2, 2021, and Emily Atkin, "Mike Rowe, Oilsplainer," *Heated,* April 13, 2021.

But the current hard pushback of the industry on antipipeline activists (such as those in Canada and the United States working to arrest the progress of Keystone XL and Enbridge's Line 3 and Line 5) and on various greening initiatives should be taken as a sign that the jig really is up.[6] Whatever else the future of energy might hold, fossil fuels will no longer constitute the planet's hegemonic source of energy; they will no longer be the basis on which the energy decisions of individuals, communities, and states are shaped or made.

But on what basis *will* energy decisions be made? Who will develop the principles and practices guiding energy transition and establish the social and political tenets of a world driven by sun and wind? Already, in the still indeterminate space between fossil fuel and renewable energies in which we now find ourselves, competing voices are at work trying to plot the direction the future will take. The renewable energy worlds imagined by the actors I've already named—McKibben, the editors of the *Globe,* and the IEA—are (unsurprisingly) distinct, as are the energy worlds mapped by other, competing voices. What remains unclear is what precisely this victory means and who will most benefit from the end of one form of energy and the beginning of another.

The future of energy remains indeterminant. The end of oil does not automatically mean the end of energy wars, environmental devastation, militaries and governments, profit and power, and wealth and poverty. The cultural imaginary that has coalesced around renewable energies can, at times, be dangerously blind-sighted. When measured against the dark, thick, exploitative, and imperialistic history of oil, it can be hard to see renewable energy as anything but its very opposite—as of necessity clean, egalitarian, peaceful, and communal. But nothing is given about the society that will emerge after oil, including the political and economic forms that will shape

6. See, e.g., Haines Eason and Emily Holden, "Solar Pushback: How U.S. Power Firms Try to Make People Pay for Going Green," *Guardian,* May 13, 2021, https://www.theguardian.com/us-news/2021/may/13/solar-power -us-utility-companies-kansas/.

it and that in turn it will shape. Solar panels on rooftops and wind
farms in fields do not guarantee that the winners of the oil era will
automatically become renewable losers. What the future will look
like depends on the success or failure of new eco-political rationalities now in the process of being created, some of which intend to
keep the sun from shining as brightly as it otherwise might.

The basic framing of the politics of energy has remained stuck
in the tired division between right and left, between those who
disavow the need for energy transition (because they disavow climate change too) and those who recognize its civilizational and
historical import and necessity. But what happens if everyone now
seems to more or less agree on what must come next? How do we
make sense of a political landscape in which a left-leaning envi-
ronmentalist (Bill McKibben); the editors of a status quo, conservative newspaper (the *Globe*); and an international agency born of
oil (the IEA) are now ready and willing to travel, hand in hand, to
the land of renewable energy? The original distinction between
right and left was an accident of the organization of parties in the
first revolutionary assembly in Paris in 1789; although it has act-
ed as a convenient political shorthand, it has long obscured both
deep and subtle historical changes in the mechanisms of political
power and the vectors of opposition to it. I want here to push past
left and right, the acceptance of transition versus its disavowal or
deferral, to suggest a different way of thinking about where we are
now with respect to the politics of climate change. Our energy and
climate futures are now being constituted in a renewable *center*
that everyone feels comfortable to inhabit—the center, that most
ideological of political spaces, which professes to be beyond ideology, where knowledge about the world is given and obvious, and
which is imagined to be navigated in an effortless and nonpartisan
way using technology and administrative reason.[7]

7. For an account of the ideology of being beyond ideology, see Slavoj
Žižek, "The Spectre of Ideology," in *Mapping Ideology,* ed. Slavoj Žižek, 1–33
(New York: Verso, 2012).

Over the next three lectures, I'll be investigating the way in which the promise of an energy center, a territory of apparent agreement about next steps to be taken, is being used to produce a "common sense" on energy transition and climate change.[8] Appeals to and articulations of common sense are among the most powerful forms of social and political power. Common sense names a quotidian consensus about the shape of the present and the direction of the future. It also understands itself to be beyond the self-interested politics of left and right; from the perspective of common sense, generally accepted practices and protocols of cultural, social, and political life are self-evident, rational, unassailable, and at times even ahistorical. But common sense is (of course) a fiction that political actors compete to shape, define, and narrate to their own ends. To do so successfully means to transform ideology into social ontology, and beliefs into realities, while making other positions and viewpoints seem vulgar, superstitious, or beside the point. This is why Roger Hallam, cofounder of Extinction Rebellion, links his group's practice of nonviolent rebellion to common sense: to legitimate it as the only possible means by which to effect climate action.[9] As we can see in the constitution of a political center in which common sense lives and thrives, many actors now feel that it has become common sense that the age of oil is over and that we should embrace renewable energy. What is far less apparent is how this common sense is being articulated, by whom, and to what ends.

8. See Hall and O'Shea, "Common-sense Neoliberalism," 8–24; Nick Srnicek and Alex Williams, "A New Common Sense," in *Inventing the Future: Postcapitalism and a World without Work*, 129–53 (New York: Verso, 2015); and Imre Szeman, "Entrepreneurship as the New Common Sense," *South Atlantic Quarterly* 114, no. 3 (2015): 471–90.

9. Roger Hallam, *Common Sense for the 21st Century: Only Nonviolent Rebellion Can Now Stop Climate Breakdown and Social Collapse* (London: Chelsea Green, 2019).

My focus here is on the arguments being made about who should be the primary agents of energy transition and climate action. Who should take the lead on developing a post-oil future and greening the environment? Now that we apparently all know these changes must happen—it's common sense, after all—who has or should have the power to make them happen? The enormous infrastructural challenge of energy transition and the scale of the environmental problems that confront us have combined to make it appear self-evident that action on energy must be led by actors with the resources to make things happen, such as government or industry. The perceived ability of these actors to shape appropriate and rational steps in energy transition and climate action is based on the social and political legitimacy they already possess. In what follows, I want to interrogate the arguments made to secure a common sense on energy transition to reveal the reality behind their rhetoric; in so doing, I hope to show how another common sense about energy futures could be created—one that can use the rhetorical power of the center to define a very different future for the energies of the sun than the one currently being shaped by the status quo.

In 2021, United Nations (UN) secretary-general António Guterres offered a forceful assessment of the Intergovernmental Panel on Climate Change's Sixth Assessment, describing it as "a file of shame, cataloguing the empty pledges that put us firmly on track toward an unliveable world."[10] His comments were directed at what continues to be seen as the greatest accomplice of climate inaction, an entity that is still, despite everything, paradoxically imagined as our greatest hope: the nation-state. And so, it is with the climate why and wherewithal of nations and nationalism that I will begin my assessment of our emerging energy common sense.

---

10. Quoted in Frank Jordans and Seth Borenstein, "UN Warns Earth 'Firmly on Track toward an Unlivable World,'" Associated Press, April 4, 2023, https://apnews.com/article/climate-united-nations-paris-europe -berlin-802ae4475c9047fb6d82ac88b37a690e.

The Temptations of Nationalism

Nation-States and Climate Change

In the politics of climate change, there is a foundational divide: either the nation-state is seen as having a fundamental role in addressing climate change or the nation-state is seen as the main problem, as the entity that prohibits or impedes action, whether by creating inadequate policies or by developing no climate change policies at all. In most cases, those committed to climate action are not fully advocates of one or the other of these positions—neither completely committed to the nation-state and looking only to nation-states as effective climate actors nor taking the position that it is entirely pointless to direct their energies to influence nation-state decision-making.

One can understand the hesitancy to figure climate action in relation to the practices of nation-states. The trauma of climate change extends far beyond borders. Carbon dioxide does not need a passport or visa to move through the atmosphere. Should there ever again be a summer in which there are no major forest fires in British Columbia—which is unlikely—smoke from California, Oregon, or Washington will still make it impossible to see the other side of the valley in south central B.C., despite the border that has been drawn across the continent at the forty-ninth parallel. The problem climate change poses is global, and nation-states are not. So, when it comes to climate action, why bother with the nation-state at all?

The answer to this question is perhaps obvious: nation-states are the only large-scale, sovereign powers that exist today. They have the capacity to *do* things. Borders are absolute fictions. But nation-states make these fictions real, erecting fences, instituting laws of transit, establishing ports of entry, and creating armies to safeguard them from external threat. Beyond making fictions into reality—a magical capacity that other political forms have found hard to mimic—nation-states also shape what takes place within their borders. Changes to structures, infrastructures, or social and economic practices that can have an impact on pollution, energy use, and transportation appear of necessity to require the work of nation-states. International agreements are not without importance or power. But nation-states are the political actors that put them into practice, whether by force or by law. International agreements do not, in the end, have the power to force sovereign nations to do anything about climate change; signatories to such agreements can decide to abandon them, even on a whim, as when the administration of U.S. president Donald Trump abandoned the Paris Agreement in 2017. Whatever one's opinions and views about nation-states, it thus seems hard to imagine any significant action on climate change and energy transition that does not involve decisions made by them—decisions rooted in the function they play within the networks of global power demarcated by property and ownership and animated by a history of violence and colonialism.

To be clear, this is not to say that other political entities do not or will not play a role in shaping our climate futures. Nations are organized around a division of power that accords discrete responsibilities to a range of internal political actors, such as cities and regions, which are also important in addressing climate change. Cities are entities that can institute curbside recycling policies or introduce bike lanes. In Canada and elsewhere, subnational entities, such as provinces and states, have constitutional responsibility to manage resources, develop power grids, and plan for energy transition. Some of the most radical and important decisions about energy transition to date have been made at these levels of governance. For example, cities and other regional governments located around the world have signed the Fossil Fuel Non-Proliferation Treaty, pledging to end extraction; still other cities (112 in total) are participants in the European Commission's Climate-Neutral and Smart Cities program, which commits them to reaching climate neutrality by 2030. Although the impact of these political actors has been significant, it is nonetheless still the nation-state that has the power to create the big carrots and even bigger sticks needed to generate climate action and energy transition with speed and at scale. Vancouver is a signatory of the aforementioned nonproliferation treaty, and the province of Quebec has indicated that it will no longer extract fossil fuels. But it is the Canadian federal government that is ultimately responsible for fulfilling the terms of the Paris Agreement. So, too, with the U.S. federal government and every other sovereign national government on the planet.

Nation-states can and do take action to address climate change. However, with some exceptions, their record to date in undertaking such changes has been miserable. The all-too-real limits of national action on climate change are evidenced by the continued increase of atmospheric CO_2. In 1958, the year the Scripps Institution of Oceanography began collecting direct measurements of the levels of atmospheric CO_2, the reading on accumulated levels stood at 315 parts per million; in May 2024, it was 427 parts per million—a 35 percent increase. The safe level of accumulated atmospheric carbon dioxide has been calculated to be 350 parts per million. Even if immediate action were taken to get the level of carbon dioxide production to zero, we would not see the latter level reached for one thousand years, according to calculations by NASA's Jet Propulsion Laboratory. In its 2022 ranking of nations based on their performance on climate change, the Climate Change Performance Index (CCPI) left the top three positions empty: no country achieved a high enough rating to fit one of those slots. In the words of the CCPI, "even if all countries were as committed as the current frontrunners, it would still not be enough to prevent dangerous climate change."[11]

The reasons for climate inaction by nation-states have been well documented by scholars and climate critics. There are vested interests that must be kept happy. Corporations, regardless of whether they express public commitments to climate change, provide marching orders (in public and in private) to members of the political establishment, reminding them of their best interests. Corporate and political elites are often drawn from the same social groups, which align their interests even without any need for backroom

---

11. Climate Change Performance Index 2022, https://ccpi.org/.

arm-twisting. In democratic nation-states, elections are always just over the horizon, and both governing and opposition parties are alert to which messages play well with publics and which don't. Affirming the reality of climate change and the need for energy transition has now become par for the course in elections worldwide, expressed as a goal by parties across the political spectrum, which can perversely create confusions about what exactly is at stake. In the 2021 Canadian federal election, for example, the Green Party and the Conservatives were distinguished not by their acceptance or disavowal of climate change but by how much they felt the nation could reduce its emissions by 2030 under 2005 levels: 60 percent for Greens versus 30 percent for Conservatives. The other parties in the election occupied an in-between territory that quickly became demarcated as reasonable and rational in comparison to these extremes.[12] This constitution of the "reasonable" or "commonsense" approach to climate has been used by governing parties to do less than needed or to defuse the need for any action whatsoever, as Canada's poor performance on climate change makes amply clear.

Yet, despite all these potential limits and barriers, and despite any evidence of substantive action to date, climate hopes continue to be placed in the actions of nation-states. This is not only due to the absence of intergovernmental sovereignties powerful enough to compel climate action; it is because nation-states are seen to possess something to which few other political entities can lay claim. Climate action has always, at its core, whether explicitly articulated or implicitly presumed, been constituted around theories of collective action and mass mobilization. Above and beyond their differences in how they understand nature or the relationship

12. See Marc Jaccard, "Assessing Climate Sincerity in the Canadian 2021 Election," *Policy Options,* September 3, 2021, https://policyoptions .irpp.org/magazines/septembe-2021/assessing-climate-sincerity-in-the -canadian-2021-election/. For a response to Jaccard, see Imre Szeman, "The Insincere Sincerity of Climate Policy in the 2021 Election," *Policy Options,* September 15, 2021, https://policyoptions.irpp.org/magazines/septembe -2021/the-insincere-sincerity-of-climate-policy-in-the-2021-election/.

between human and nonhuman, all forms of climate action lay claim to a theory of the political and to the mechanisms by which political change occurs. Peaceful protest? Violence? Mass demonstrations? Voting? Pamphleteering? Advertising? Twitter campaigns? Petitions? All of these? None? One of the oft-repeated frustrations about effective climate action emerges from the apparent limits of all such forms of environmental communication to generate significant action. By contrast, nation-states possess a capacity to suture together belief and action in a single process. They have a proven method for compelling their citizens to act quickly, en masse, and collectively: *nationalism*.

An environmental nationalism promises to accomplish several things at once. It would resolve climate knowledge into climate practice, and do so in relation to a common project, whose import—national *and* international—would be foregrounded for populations now seen as either disinterested in climate change or intimidated by what needs to be done to address an unprecedented social, political, and ecological challenge. The various appeals made to nation-states to carry out Green New Deals, or to mobilize what activist Seth Klein has called "a good war" against climate change, each speaks to this capacity of the nation-state, if in varying ways and to different degrees.[13] The view that nation-states could make a difference—and fast—if they were to treat carbon dioxide as an enemy is not uncommon; indeed, many have come to see it as the only way forward, even if one has to hold one's nose to avoid the stink of concession to politics as now practiced.

Climate action and energy transition likely must tarry with the nation-state in some way, given today's organization of sovereign power and the necessity for immediate action. But need they do so via nationalism? And just what does it mean to do so? Before I answer these questions, it's necessary to take a slight detour into

---

13. Seth Klein, *A Good War: Mobilizing Canada for the Climate Emergency* (Toronto, Ont.: ECW Press, 2020).

the contemporary landscape of nationalism; how this has been constituted has a significant impact on whether nationalism can in fact accomplish substantive action on climate change and energy transition or if it constitutes little more than a ruse—a promise of action that will never in fact be taken but that reaffirms the position of nation-states at the heart of the story of climate change, with all this entails for alternative and more radical visions of the future of the sun.

Bad Nationalism, Good Nationalism

Climate change isn't the only place where nationalism is being discussed today as a force that might save us from our worst tendencies. In books and articles written recently by a number of scholars and critics—I am thinking here, for instance, of Anatol Lieven, Yael Tamir, and the prophet of the end of history, Francis Fukuyama—a renewed nationalism is being proffered as the solution to the scourge of populist totalitarianism.[14] The way this nationalist politics is being articulated has consequences for how—and whether—nationalism might actually work as a mechanism to achieve energy transition and address climate change.

The idea that nationalism is the way to confront populism and totalitarianism can't help but come as a surprise. After all, isn't nationalism precisely the *cause* rather than the solution to national and international woes? When I think of present-day nationalisms, I think of Recep Tayyip Erdoğan's Turkey, Narendra Modi's India, or Viktor Orbán's Hungary. A range of forces and practices define contemporary populist nationalism, including the significant role that social media plays in animating ideology and suppressing dissent. But at heart, contemporary totalitarianism is constituted in

14. Francis Fukuyama, *Liberalism and Its Discontents* (New York: Farrar, Straus, and Giroux, 2022); Anatol Lieven, *Climate Change and the Nation State: The Case for Nationalism in a Warming World* (Oxford: Oxford University Press, 2020); Yael Tamir, *Why Nationalism* (Princeton, N.J.: Princeton University Press, 2019).

much the same way Hannah Arendt describes in *The Origins of Totalitarianism*: around inclusion and exclusion, the promise offered to some of absolute belonging and identification—often as a salve for social isolation and loneliness—in relation to the supposed threat others pose because of their race, ethnicity, or political difference.[15] If nationalism is so important to illiberal nations, how exactly is the former also supposed to challenge the principles on which the latter are constituted?

The answer that these critics offer is surprisingly simple—which is not to say that it's satisfactory. Lieven argues that while there might be a growing number of "bad" nationalisms—to use a shorthand—this does not ipso facto rule out the possibility of establishing *good* nationalisms. Bad nationalisms are defined by prohibitions and surveillance and by the manner in which elites use exclusions and violence to gain and then maintain power (consider Modi vs. Muslims, Erdoğan vs. Kurds, Orbán vs. Jews and their coconspirators). But against this form of nationalism, Fukuyama, Tamir, Lieven, and others argue that good nationalisms are also possible. Such nationalisms are defined by modes of being-in-relation-together framed around values that have long defined liberalism: freedom, the rule of law, tolerance of difference, pluralism, and respect for and a commitment to individual rights. Contra populist nationalism, good nationalism insists that universal rights and freedoms be constitutive of the political and civic life of all nations. In contrast to the illiberalism of populist nationalism, its advocates see good nationalism as shaped according to universal liberal values, such as those captured in the language of the UN's Universal Declaration of Human Rights.

A conceptual and practical tension seems to be arising from the conjunction of liberal universalism and articulations of good nationalisms. This is a tension, it must be said, that mirrors commitments

---

15. Hannah Arendt, *The Origins of Totalitarianism* (1951; repr., New York: Penguin Classics, 2017).

to climate change made via *international* accords, such as the Paris Agreement, that must nevertheless be fulfilled by nation-states; it is also a tension foundational to liberalism as such. One must ask, if the values of liberalism are understood to be universal, an expression of a core set of political values developed in relation to a belief in liberty and equality for all, why do they need to be expressed via nationalisms, which are of necessity particular and exclusionary? Why turn to nationalism and not to postnational forms of belonging, which might better challenge and contest the parochialism and divisions generated and maintained by national borders? In "A Country of Their Own: Liberalism Needs the Nation," an essay in *Foreign Affairs* that offers a condensed version of the argument he makes in his 2022 book, *Liberalism and Its Discontents,* Fukuyama provides a direct answer to this question: nation-states are the only form of sovereign power in the world; they are the only actors able to enforce the rule of law and create the conditions to guarantee liberal universalisms.[16] It is unavoidable: nations must come first. Bad nationalisms also place the nation first, using sovereign power to make determinations about (for example) freedom of speech and assembly in a manner that suppresses liberalism and secures belief in and commitment to illiberal ideologies. What thus makes a nationalism good comes down to a single thing: it builds collective belonging and common cause around universal values, around inclusions instead of exclusions, pluralism instead of homogeneity, and with a commitment to the rule of law and respect for individual rights.

When it comes to *how* nationalism operates as a form of collective belonging linked to a common project, the difference between the bad nationalisms now in existence in all too many places on the globe and the good nationalisms advocated by Fukuyama and by Tamir in *Why Nationalism* is a very fine one. It hinges on little

16. Francis Fukuyama, "A Country of Their Own: Liberalism Needs the Nation," *Foreign Affairs,* May/June 2022, https://www.foreignaffairs .com/articles/ukraine/2022-04-01/francis-fukuyama-liberalism-country.

more than a vision of the character of belonging articulated within nationalism—either a nationalism shaped around the rejection of liberal humanism or one that embraces it. The difference advocated by these critics is taxonomic more than it is political: there are two types of nationalism, and it would be better to have one (a nonchauvinistic one) instead of the other (defined by the chauvinisms of populism). Fukuyama fully understands the difficulty of making a case for liberal nationalism, especially given the obvious attractions of populist nationalisms: the latter offer individuals an easy way to make sense of an increasingly complex world, despite the consequences for communities within nations and the need to make common cause by limiting who belongs to the common, often by the threat of violence or its actual use. The political task liberals face today is to newly articulate its values, attractions, and importance to all those left behind, to reconstitute liberal societies by ensuring that everyone can thrive within them. But precisely how this is to be done is never elaborated, which makes the arguments for good nationalism little more than subject matter for *New York Times* op-eds and essays in *Foreign Policy*.

The arguments by Fukuyama and Tamir, and other recent appeals to a renewed liberalism, are shaped around a mea culpa: post-1989, the attractions of globalization made liberals insensate to the experiences and struggles of those whose lives were negatively impacted by neoliberalism. The fact that this mea culpa has appeared *three decades* after Fukuyama's infamous proclamation of "the end of history" suggests that these thinkers have failed to pay any attention to social and political movements and academic analyses that have relentlessly drawn attention to the traumas produced by globalization over this time period.[17] (We will witness other such mea culpas over the course of these lectures; they are a powerful rhetorical tool.) Tamir suggests that liberals became mobile and

17. Francis Fukuyama, *The End of History and the Last Man* (1992; repr., New York: Free Press, 2006).

forgot about the immobile classes; they couldn't see them from the cruising altitude of their corporate jets. Twenty-five years ago, sociologist Zygmunt Bauman had already made much the same point: the advent of globalization had divided humanity into tourists and vagabonds, the first group able to move easily across borders, the latter moving only when forced to do so, due to poverty, war, or climate destruction.[18] The arrogance developed by wealthy globalists is captured in Hillary Clinton's description of Trump supporters during the 2016 U.S. presidential election campaign as a "basket of deplorables," half of whom were "racist, sexist, homophobic, xenophobic, Islamophobic."[19] In *Why Nationalism*—tellingly, the title has no question mark—Tamir writes, "Members of the middle classes . . . lost trust in the ruling liberal elites and questioned their willingness to protect the interests of the different social classes."[20] It is this trust that Fukuyama, Tamir, and other liberals now want to regain, by striving to protect (they claim) the class of immobiles and vagabonds from the threats and dangers they now daily encounter. Whether they have any similar ambition to help vagabonds become more like tourists, or to reduce the air miles accumulated by liberal elites, is far less certain.

I've taken the time to elaborate these recent arguments being made on behalf of a renewed, postglobal liberalism for two reasons. First, the liberal nationalism being articulated here is one whose success depends on its ability to better attune itself to conditions on the ground—that is, to class divisions exacerbated by global processes within nation-states. Despite the desire of liberal nationalism to highlight universal values in the governance of individual nations, as a political project, the reinvigoration of good nationalisms is today

18. Zygmunt Bauman, *Globalization: The Human Consequences* (London: John Wiley, 1998).

19. John Cassidy, "Hillary Clinton's 'Basket of Deplorables' Gaffe," *New Yorker,* September 11, 2016, https://www.newyorker.com/news/john -cassidy/hillary-clintons-basket-of-deplorables-gaffe.

20. Tamir, *Why Nationalism,* xliii–xliv.

being narrated as necessarily *anti-internationalist*. The political work of countering populist nationalisms and rebuilding trust in elites demands attention to local, national projects—job growth, the strengthening and expansion of social services, and improvements to health and education—designed to motivate the class of deplorables to switch ideological teams. Given this emphasis, it is perhaps unsurprising that this version of liberal nationalism appears to have little interest in climate change as a problem worth worrying about, at least in the short term. It is telling that neither Fukuyama nor Tamir mentions climate change in his or her book-length argument for why the world needs good nationalisms.

This brings me to my second point. The limits of liberal nationalism as an environmental project adhere in the necessity for climate action to take place within the *right kind* of nationalisms—good nationalisms committed to a universal project. Liberal nationalism always imagines itself as a universal project, even if this is a universal conducted within the borders of a given nation. So, too, with liberal nationalisms committed to climate change: the fight against climate change is necessarily global (remember, CO_2 and smoke from forest fires don't obey borders). The promise of liberal nationalism—the argument that underlies many of the beliefs that the existing nation-state system is sufficient to address climate change—is that the sacrifices of citizens for the good of their national communities are at the same time acts for the good of the globe. This fantasy is embodied in the Paris Agreement, which pretends to act as something like an environmental constitution that aggregates discrete nationalisms into a single global project. But an effective transmutation of the national into the global depends on the presence of good nationalisms everywhere in the world. It is unlikely that such a network can ever be completed—and in the absence of such a network, any given liberal nation-state is provided with a convenient argument for the lack of true action on energy transition and climate change: the local can become global only once the whole planet is liberal, and that is far from where we find ourselves today. The limit for effective action on climate change thus isn't liberal nationalism as a project at the level of the nation but the absence of a universal system of good nationalism. Or so the argument goes.

A world of good nationalisms doesn't exist, nor is there anything to suggest that even the best renewed liberal nationalisms—ones now attentive to the classes they had so recently disavowed—would be especially attuned to the challenges that climate change poses. There is certainly no evidence that existing liberal nation-states have been more effective in addressing climate change than populist or totalitarian ones. On the CCPI, Brazil ranks above New Zealand, Turkey ranks above Japan and Ireland, and almost every nation ranks above Canada (which sits sixty-first on a scale with sixty-four places).[21] Troublingly, populist nationalisms are rapidly incorporating environmental action into their narratives, legitimating racial and ethnic exclusions as necessary to defer further degradation of the national soil.[22] Despite its attractions and promises, it is hard to see nationalism as a mechanism for resolving the political challenges of producing action on climate change. And yet, the common sense of environmental action that persists is that the nation-state remains the single most powerful body to secure energy transition and to undertake climate action, despite how slowly it might actually be able to do so.

Time, Nationalism, and Climate Change

I've noted the absence of climate change as a topic in the recent books on liberalism and nationalism by Fukuyama and Tamir. By contrast, there is a book that explicitly endeavors to connect the dots between the politics of nation-states and action on climate change: Anatol Lieven's *Climate Change and the Nation State: The Case for Nationalism in a Warming World*. Many of the arguments he makes on behalf of nationalism echo those I've outlined herein. Lieven,

21. Climate Change Performance Index 2022, https://ccpi.org/.
22. See Sam Moore and Alex Roberts, *The Rise of Ecofascism: Climate Change and the Far Right* (Cambridge: Polity Press, 2022).

too, makes the claim that there is a need for "good" nationalisms and argues that today's troubling nationalisms don't rule out the possibility that good variants might be created. The force of his argument, however, comes from a different place. Lieven strongly insists that nationalism is the *only* means by which to address climate change. The nation-state is the only sovereign actor capable of producing mass change, and nationalism is a proven mechanism by which states can mobilize their populations to act. As such, there is no possibility for effective energy transition or action on climate unless nationalism is mobilized for this purpose.

The subtitle of Lieven's book suggests that he is arguing for a position by outlining its rationale; this is, after all, what it means to make a case. In the end, however, the argument he proffers is premised less on the articulation of the specific effectiveness of nationalism than on the elaboration of a position by means of negation: a relentless criticism of all those who might be wary of or even directly opposed to the nation-state as climate actor and nationalism as its tool of action. Repeatedly, Lieven positions the nation-state as the only force capable of undertaking climate action. He is ferocious in his attacks on left political theorists like Michael Hardt and Antonio Negri or philosophers like Alain Badiou, all of whom he sees as presenting models of political and social change that are, in the end, little more than speculative and fantastical and so not to be countenanced. He is equally critical of environmental critics and advocates like Naomi Klein and Astra Taylor—"committed opponents of capitalism tout court"—who evoke the need for transnational approaches to addressing climate change and make appeals for the end of capitalism.[23] For Lieven, these critics articulate no mechanism by which to generate public consent for policies about climate action. The necessity of nation-states and nationalisms is affirmed largely because no other feasible political approaches to climate change are available.

23. Lieven, *Climate Change and the Nation State*, 93.

As luck would have it, Lieven sees nationalism as a perfect substitute for the political commitments articulated by socialists, Marxists, and other radical environmentalists (though it must be remarked that less radical ones, such as McKibben, also want to rein in capitalism). This is perhaps especially so in relation to time. Lieven writes, "Apart from nationalism's legitimization of painful reforms in general, what especially brings nationalism and ecological thinking together is the capacity of both to demand sacrifices on the part of existing people for the sake of future generations."[24] This understanding of the place of time in nationalism has a long history. It can be found in Edmund Burke's infamous description of the social contract as "a partnership not only between those who are living, but between those who are living, those who are dead, and those who are yet to be born."[25] It can be found, too, in Ernest Renan's definitive 1882 essay on nationalism, "What Is a Nation?" Renan's answer, which also comes about through a negative process, via eliminating other ways of being together (e.g., tribes, city-states, "great agglomerations of men"), is that

> a nation is . . . a great solidarity constituted by the feeling of sacrifices made and those that one is still disposed to make. It presupposes a past but is reiterated in the present by a tangible fact: consent, the clearly expressed desire to continue a common life.[26]

Lieven sees clear parallels between the time of nationalism and appeals commonly made within environmentalism to safeguard future generations through present-day sacrifices. Indeed, he sees an overlap between the time definitive of nationalism and the "alleged 'seven generations' thinking of the Iroquois people of North America."[27] Given such shared commitments to time, ones that also

24. Lieven, 76.

25. Edmund Burke, *Reflections on the Revolution in France,* ed. Frank Turner (1790; repr., New Haven, Conn.: Yale University Press, 2003), 82.

26. Ernest Renan, "What Is a Nation?," in *Qu'est-ce qu'une nation?,* trans. Ethan Rundell (Paris: Presses-Pocket, 1992).

27. Lieven, *Climate Change and the Nation State,* 80.

produce forms of common cause and self-sacrifice akin to war, Lieven argues that there should be nothing for those committed to climate change to worry about: nationalism will take care of it all.

Despite what might appear to be structural similarities, commitments to the future (and to the past) need not affirm nationalism or the nation-state in any way. Indigenous understandings of the past and present don't emerge from, nor are they imagined in relation to, nationalism as articulated by Burke and Renan; the claim Lieven makes that equates the former to the latter is offensive in the extreme. Nor can the political self-understanding of movements like Fridays for Future ("We are fighting for our future and for our children's future") be categorized as something akin to nationalism; whatever else Greta Thunberg might be doing, she's not gearing up for war or vying for the presidency of a proto-nation made up of youths. As in the case of the arguments he makes against those he deems to be socialists, Lieven's intent is to rule out alternative forms of climate action by affirming the rationality of the status quo. Or rather, it is to make claims about circumstances not yet in place but that he treats as status quo—a coming good nationalism that stands ready to mobilize populations via a good Green New Deal.

And so, we return again to the notion of a good nationalism. What defines this nationalism is that it retains all of the constitutive collective power attendant to it—that is, charging subjects to act against their own will for the good of the nation, whether this external threat is the result of geopolitics or climate change—while safeguarding against potential antiliberal extremes. In the end, Lieven's argument against (what he sees as) the speculative and utopian viewpoints that Klein and others have articulated comes down to the speculative and utopian elaboration of a prospective good nationalism. He writes,

> Sixty years on, unconstrained free-market capitalism has once again been running amok for a generation, with disastrous results; and many socialists in the West have once again abandoned loyalty to their nations in favor of a return to fantasies of a borderless progres-

sive world guided by themselves—and looming in the background, unaddressed, is the threat of climate change to all existing states. *The task then is to develop a new version of social imperialism without the imperialism, racism, eugenics, and militarism.*[28]

And again:

A future US president who hopes to push through a Green New Deal *will need to combine Obama's appeal to core US traditions with much greater radicalism, backed by a much stronger appeal to nationalism.*[29]

Radicalism without too much radicalism; imperialism without imperialism; a nationalism drained of all the dangers of nationalism; a status quo liberal capitalism inflected by a kinder, gentler ethos: by adopting a supposedly realistic, commonsense position about what can change and what can't, Lieven argues that a good nationalism can secure future possibilities in a manner that socialist internationalisms supposedly cannot.

For reasons I've named at the beginning of this lecture, there is an understandable temptation to concede climate action to the nation-state and to turn to nationalism as an effective mechanism of individual and infrastructural change. There are status quo versions of this—Lieven's being one example—and more radical ones as well: Andreas Malm's "war communism" is one example; another is Chantal Mouffe's argument for a "left populism"; still another is Fredric Jameson's articulation of a citizen army, which bizarrely enough evokes Lieven's own account of the environmental benefits that might be produced by required national service.[30] But to this one can't help but raise a simple point: *there is no good nationalism.* There is no harm in arguing for models of sociopolitical structures

28. Lieven, 98, my emphasis.

29. Lieven, 102, my emphasis.

30. Lieven, 90; Andreas Malm, *Corona, Climate, Chronic Emergency: War Communism in the Twenty-First Century* (New York: Verso, 2020); Chantal Mouffe, *For a Left Populism* (New York: Verso, 2019); Fredric Jameson, *An American Utopia: Dual Power and the Universal Army* (New York: Verso, 2016).

that might be better than the ones we now occupy. When such models are taken as evidence of the utter infeasibility of other political positions, we are witnessing an attempt to take full control of conceptual space within which it might be possible to contest political common sense. If their arguments are pushed to the wall, the trump card that Lieven and other liberals play is sovereignty: only nation-states have it. Environmental activists and others who recognize the importance of rapid climate action and energy transition need to challenge the legitimacy of this card trick; they may even be able to do so via the language of nationalism, not to argue for their own variant of a good nationalism, but to articulate new ways of being in relation to one another, even within the system of nation-state sovereignty. (I'll say more about how they might do so in lecture 3.)

My aim in this first lecture has been to show the distinct and complex ways in which the nation-state and nationalism has been configured today as a site of climate action. Despite the failure of nation-states to take significant action on climate change—as significant as needed, given the size of the threat—strong claims continue to be made about the key role of nation-states in fighting climate change; they are imagined to be the only real climate actors in town, whether or not one might want this to be the case. Yet this role is effective only to the degree that it originates from within the terms of what the critics at whom I've been looking have envisioned as a good nationalism. Conceptually, such an imagined nationalism does two things at once, both of which secure the status quo organization of power. Good nationalism guarantees the collective internationalism of nation-states by virtue of the fact that such states are liberal, which in turn rules out the worst tendencies of nationalism witnessed in contemporary, "bad," populist nationalisms, because the former have a different ideological valence; that is, they're good. This nested set of tautologies, almost impossible to parse, should act as a warning. Even if one were to concede the reality of the sovereign capacities of nation-states, and so, too, their potential for real climate action, it is crucial that we not concede that climate action has to occur (or can only occur) through nation-states. If nothing else, articulations of international climate solutions, which Lieven rules out as inconsequential, are necessary to unsettle the status quo fantasy of the promise of eco-nationalism as a last-ditch way to take on climate change. Waiting for good nationalisms to take charge will only lead to more of the same: not climate action at scale and with speed but too little, too late. The environmental and social consequences of climate change are too serious to concede action to a self-satisfied liberalism that imagines itself as the only game in town.

Lecture 2. The Life and Times of Bill Gates, Eco-Warrior

The Birth of Entrepreneurs

In my last lecture, I examined the environmental hopes placed in the nation-state, which has come to be (and likely always already was) imagined as the key actor in the struggle to address climate change. It's obvious why many would see the nation-state in this way. Nation-states are imagined as having the resources necessary to take substantive action on climate and to do so at the scale and with the speed demanded by the crisis at hand. My examination of the way in which discourses of nations and nationalism are being mobilized for climate action today suggests that, unfortunately, nations are unlikely to play the role one might have hoped. Appeals to *nationalism* as a mechanism to get people to act on climate—often imagined as akin to mobilizing them for war—come at a time when populist nationalisms have unnerved the belief of liberal nationalism in its own powers. The only guarantee that nation-states will use nationalism correctly as a mechanism for climate change is that nations understand and undertake their mission in the right way—that is, via the values and qualities typically associated with liberalism. The lack of significant action to date by liberal nation-states; the questionable, easy equation of liberal capitalism with "good" nationalism; and the present crisis of nationalism all suggest that climate hopes placed in the nation-state are unlikely to be rewarded with substantive and effective climate action.

If not the nation-state, then what? If nations are unlikely to generate new eco-possibilities, either on their own or due to citizen pressure, where are we left? What other arguments have been made to establish control over narratives of energy transition, climate action, and the shape of our renewable futures? Are any other actors able to instigate impactful environmental change with the speed and at the scale needed to make a real difference? Who are they, and why might they be interested in doing so?

To answer these questions, one must dig into the historical damp and humus from which contemporary eco-actors have grown and the conditions of possibility that would permit them to continue to flourish. Accounts of the recent fate and state of nations (and the nationalisms associated with them) are inevitably told through narratives of their rise and fall. The period following World War II is characterized by triumphant nation-states competing against one another to establish the global hegemony of individual nations (e.g., the United States, the USSR) or groups of nations (e.g., the North Atlantic Treaty Organization, the Warsaw Pact); nation-states expressing a commitment to internationalism after the trauma of war (e.g., the UN); postcolonial nationalisms (e.g., India, Pakistan, Nigeria, the Organisation of African Unity), which challenged the physical and psychological violence of colonialism to create new collectives, some more successfully than others; and social democratic nation-states (in whatever form and to whatever degree) that were the result of decades of workers' struggles to create polities attentive to the needs of all their denizens. These different modalities of nationalisms—and still others I haven't named—sometimes supported and reinforced one another (e.g., the Cold War and postwar Keynesianism; Cold War competition and the birth of postcolonial nation-states) and sometimes threatened each other's viability (e.g., exuberant postcolonialisms put the lie to the Cold War rationality that played a role in bringing them about).

And so, the rise—but what of the fall? The way in which the upended chessboard of the Cold War system was put back together is captured in a single word: globalization. But chess played without an opposition, with pieces of a single color, turns out to be no game at all: globalization suggests that following the Cold War, nation-states are no longer necessary. This promise of a postnational world was, we know, a ruse.[1] Even if borders disappeared for the movement of capital and goods, they only ever allowed some bodies to move through passport checkpoints with the freedom enjoyed by financial markets. What is certain about globalization is that the deterioration of the postwar system of competing nation-states generated possibilities for capital that had not previously existed. Neoliberalism—the ideology at the heart of the global era—predates the birth date of globalization; after a successful test run in Chile in the 1970s, it was already being put into play by the governments of Ronald Reagan, Margaret Thatcher, and Brian Mulroney in the 1980s. But neoliberalism could become hegemonic only because of the openings that globalization created for crafting postnational political subjects shaped by new rationalities (or if not entirely new ones, older ones put into play with more intensity). In place of the citizen—a subject with obligations to the nation-state, but also vice versa—was born the entrepreneurial subject of neoliberalism, to which nation-states owe no obligation other than the assurance that they will get out of the way to allow this new subject to do its work.[2]

1. For a review of the limits of globalization, see Imre Szeman, "Globalization," in *The Johns Hopkins Guide to Literary Theory and Criticism,* 2nd ed., ed. Michael Groden, Martin Kreiswirth, and Imre Szeman, 458–65 (Baltimore: Johns Hopkins University Press, 2003).

2. With respect to neoliberalism and subjectivity, see, e.g., Wendy Brown, *Undoing the Demos: Neoliberalism's Stealth Revolution* (Boston: Zone Books, 2017); Jonathan Crary, *24/7: Late Capitalism and the Ends of Sleep* (New York: Verso, 2014); Pierre Dardot and Christian Laval, *The New Way of the World: On Neoliberal Society* (New York: Verso, 2017); Michel Foucault, *The Birth of Biopolitics: Lectures at the Collège de France, 1978–1979,* ed. Michael Senellart, trans. Graham Burchell (New York: Picador, 2010); and Annie McClanahan, "Becoming Non-economic: Human Capital Theory and Wendy Brown's *Undoing the Demos," Theory and Event* 20, no. 2 (2017): 512.

Writing about entrepreneurial subjects has focused in the main on the tragic deformation of social and political life that their existence represents.[3] Where once there was a space of democratic deliberation and decision-making (to whatever degree and extent; ruses abound in political life) that took place in a sphere deliberately set aside from the market, all aspects of social life had become subject to the logic of economics. This was a political project built with a specific intent, however messily and imprecisely this may have been (and continues to be) executed: to limit the capacity and the political necessity of the state to provide social goods.[4] This withdrawal of aspects of the state from public life has happened throughout the world, with different consequences in each polity where neoliberalism has taken hold.

Although a great deal has been written about neoliberalism and its impact on the subjects who have had to use its ideologies and rhetoric as the framework for their existence, less has been written about another figure that neoliberalism has produced. This is what I will call *meta-entrepreneurs,* composing that class of celebrity billionaires whose success and power both affirm the legitimacy of neoliberalism and provide entrepreneurial citizen-subjects with evidence of what they can accomplish if they just try hard enough. As the grand political end product of neoliberalism, meta-entrepreneurs possess powers that states do not, and not only because they have gained social and economic dominance in the wake of the withdrawal of the state. Their powers arise from their ability (sometimes real, sometimes imagined) to make things happen. States are slow to act, even now seeking and requiring consent from citizens before they can enact grand social transformations, for ex-

3. See, e.g., David Chandler and Julian Reid, *The Neoliberal Subject: Resilience, Adaptation and Vulnerability* (Lanham, Md.: Rowman and Littlefield, 2016), and Rosalind Gill and Christina Scharff, eds., *New Femininities: Postfeminism, Neoliberalism and Subjectivity* (London: Palgrave Macmillan, 2018).

4. The now classic articulation of this can be found in David Harvey, *A Brief History of Neoliberalism* (Oxford: Oxford University Press, 2007).

ample, the establishment of a society committed to climate war. By contrast, meta-entrepreneurs, figures connected to companies with monikers that sound as if they are taken from science fiction novels (e.g., Alphabet, Meta, Apple), can largely do what they want: there remains remarkable little state oversight on what these companies do given their astonishing social, political, and economic power.[5]

To date, only a few meta-entrepreneurs have concerned themselves directly with environmental issues. Among these, one figure is playing an outsized role in defining what our energy and environmental futures will look like: Microsoft cofounder Bill Gates. Given the contemporary neoliberal political landscape and the weakness of the liberal nation-state, how Gates imagines addressing climate change and engendering energy transition needs to be taken seriously. Gates's postnational account of best-case climate scenarios and views on what to do and what comes next is giving shape to an environmental common sense that may well prove to be difficult to challenge or unnerve.

Let me say it again: if not the nation-state, then what? For some, the answer is the socially concerned meta-entrepreneur, exemplified par excellence by Bill Gates. But before we get to Gates, we need to grapple with the ways in which one of the most important dimensions of climate change—energy—has been sutured into the imaginaries of the ruling class, past and present.

From Steam Fetishism to Solar Fetishism, 1825–2025

The transition from fossil fuels to renewable energies constitutes the second modern energy transition, the first being the shift that took place in response to the mass adoption of fossil fuels. That there are historical differences between the two transitions is ob-

5. See Shoshana Zuboff, *The Age of Surveillance Capitalism: The Fight for a Human Future at the New Frontier of Power* (New York: Public Affairs, 2020). I would argue that this remains the case, even given laws instituted in the European Union and elsewhere intended to provide some oversight of companies like the ones just mentioned.

vious enough; the two hundred years between 1825 and 2025 are different enough to render weak even broad analogies. For a start, the circumstances of the present transition have been shaped profoundly by the earlier one; for this reason alone, to see the transition *to* and *from* fossil fuels as being an energy transition in the same sense is a category mistake. Nonetheless, understanding the structural and ideological forces guiding the first transition is a useful way of grasping just what's different this time around—and of understanding, too, the broad outlines of the energy worldview within which Gates and other meta-entrepreneurs operate.

Malm's *Fossil Capital* provides one of the best accounts of the first energy transition.[6] The book endeavors to offer a counternarrative of the birth of energy by mounting a protracted challenge to dominant narratives that position the steam engine as the protagonist of energy history and, indeed, place it at the origin of modernity itself. The arrival of the steam engine in 1776 (so perfectly aligned with the American Revolution that one can't help but think someone impish is pulling the levers of history behind the scenes), so the story goes, set into motion a series of historical developments that led inevitably to the creation of fossil-fueled modernity. Malm mounts a detailed, multifaced challenge to this narrative, anchored by a fact of history that is impossible to refute. He notes that the steam engine—which is to say, the use of coal as a source of energy—did not become the prominent source of energy in the cotton industry (a primary site of energy use in the early nineteenth century) until well after its invention. Steam was adopted in fits and starts by factory owners and industrialists mostly after the economic crash of 1825. It wasn't until the middle of the nineteenth century that coal truly became the primary source of energy—some *seventy-five years* after the date most often marked as the beginning of the fossil fuel era.[7]

6. Andreas Malm, *Fossil Capital: The Rise of Steam Power and the Roots of Global Warming* (New York: Verso, 2016).

7. Analyses from other perspectives offer a similar account of the slow speed of transition. Roger Fouquet writes, "Since the Industrial Revolution, it has taken, on average, nearly fifty years for sector-specific energy transitions

It's easy to see why steam was so slow to be adopted. The original sources of energy used for cotton production—water and wind—were free, whereas coal was not. Water also produced far more power than early steam engines. A complex set of factors resulted in the adoption of steam power as the predominant form of energy. At heart, however, steam was adopted despite its greater cost because it allowed factory owners to better manage the labor necessary for expanded cotton production, especially in the wake of the powerful labor movements engendered by the economic crash. Unlike water, coal could be transported to sites of production in cities, where labor was ample and competition for jobs fierce, and where striking workers could be replaced by other workers desperate for income, or by children and women, whose labor proved easier to exploit.[8] Although transporting coal was expensive, bringing coal to factories also eliminated the need to transport labor to water mills, where manufacturers would have to bear the costs of housing and feeding them, often with disastrous results for production.[9] Far from being the protagonist in the story of fossil-fueled life, Malm relegates the steam engine to being little more than a plot device—important to moving along the story of fossil fuel transition, but hardly the main actor.

Malm's exploration of the forces and dynamics guiding the transition from water to coal—from what he terms the energy of *flow* to *stock*—stands in contrast to the solar transition currently under way—which is a return move, from *stock* back to *flow*. Whatever else one might say about the transition from water to coal, it was never conceived as a sociopolitical project *deliberately and directly* focused on transfiguring energy use to a determinate end in the manner of current energy transition—that is, a project focused on producing

---

(i.e., the diffusion of energy sources and technologies) to unfold in the United Kingdom." Fouquet, "Historical Energy Transitions: Speed, Prices and System Transformation," *Energy Research and Social Sciences* 22 (2016): 7.

8. See Malm, *Fossil Capital,* 121–64.

9. Malm, 165–93.

energy differently and doing so on a global level and in short order. It is also hard to see the current transition as emerging from
the need to manage labor struggles.[10] The struggle to adopt solar
power isn't in the main the outcome of labor problems connected
with the use of fossil fuels, due in part to the general suppression
of the labor movement in the neoliberal era. Nor is the shift to solar
everywhere and only an elite project dedicated to ensuring that
industrial processes work without impediments or disruptions.
Comparisons between the first and second transitions turn up few
similarities but many differences.

The forces driving the current energy transition are to be found
elsewhere. The causes are as complex as those mapped by Malm
with respect to the first transition. Energy cost is once again part
of the story. Coal was adopted despite being more expensive than
its alternative. By comparison to fossil fuels, solar power (taken
here as a synecdoche for all renewables) is cheap once capital is
invested to develop and install it, and so one might imagine capital racing to adopt solar power to lower production costs. Several
things stand in the way of the rapid adoption of solar power. These
include barriers erected by industry and the weight of the physical
and social infrastructures developed over two centuries around
the use of fossil fuels.[11] Solar power also threatens low returns on
investment. The inversion here is important to note. Capitalists
were once dragging their feet to leave flow behind, until the benefits
of its use in production became evident. Now some are dragging
their feet to return to it, despite potential cost advantages and the

10. On labor and energy, see Huber, *Climate Change as Class War,* and
Timothy Mitchell, *Carbon Democracy: Political Power in the Age of Oil* (New
York: Verso, 2013).

11. On the significance of infrastructure for energy, see Jean-Paul
Deléage, Jean-Claude Debeir, and Daniel Hémery, *In the Servitude of Power:
Energy and Civilization through the Ages* (1991; repr., London: Zed Books,
2021); Jeff Diamanti, *Climate and Capital in the Age of Petroleum: Locating
Terminal Landscapes* (London: Bloomsbury, 2023); and Keller Easterling,
Extrastatecraft: The Power of Infrastructure Space (New York: Verso, 2014).

environmental consequences of not doing so—and the growing import of being seen by consumers as on the right (i.e., green) side of energy history.[12] The reasons have less to do with labor than they do with the sheer amount of power provided by fossil fuels and the desire of every industry to squeeze as much revenue as possible from its sites of extraction.

Even if slowly to date, the adoption of renewables is now beginning to accelerate, and for reasons wholly absent in the transition Malm examines.[13] In addition to the potential for cost savings, solar transition has been mobilized and sustained—cynically at times, deeply felt at other moments—by ethical and eschatological considerations, by a sense of responsibility to future generations and other inhabitants of the planet, and by the goal of deferring or delaying the end times that constitute the ultimate threat of climate change.[14] Commitments to ethics and reflections on eschatology might seem far afield from the core principle of capital: the accumulation of profit. But remember, if using fossil fuels didn't generate carbon dioxide, and carbon dioxide didn't make fundamental changes to

12. The 2022 U.S. Inflation Reduction Act (IRA) foregrounds the need for government to invest in renewables to make them profitable. See Pippa Stevens, "First Solar Announces New U.S. Panel Factory Following the Inflation Reduction Act," CNBC, August 30, 2022, https://www.cnbc.com/2022/08/30/first-solar-to-build-new-panel-factory-following-inflation-reduction-act.html. Matt Huber offers a very different read on the imperatives driving the IRA. See Huber, "Mish-Mash Ecologism," *New Left Review Sidecar*, August 18, 2022, https://newleftreview.org/sidecar/posts/mish-mash-ecologism.

13. In its 2022 World Outlook Report, the International Energy Agency noted that for every US$1 investment spent globally on fossil fuels, US$1.50 is now spent on clean energy. The IEA scenarios for future energy investments show this difference growing by as much as 1:5 by 2030. See https://www.iea.org/reports/world-energy-outlook-2022/executive-summary.

14. See Allan Stoekl, "'After the Sublime,' after the Apocalypse: Two Versions of Sustainability in Light of Climate Change," *Diacritics* 41, no. 3 (2013): 40–57.

the planet's atmosphere, it's unlikely there would be much, if any, interest in a transition from stock back to flow. There would be no climate problem to address and no crisis to manage. Might the biggest distinction between the first and second transitions—between the shift to fossil fuels and the shift away from them, between the one Malm maps and the one in which Gates participates—be the introduction of a set of values into capitalism that one would never have expected to find there, values that some have thought might unnerve the accumulation of profit?[15]

In a crucial chapter of *Fossil Capital,* Malm describes the process by which steam became an essential component of nineteenth-century bourgeois ideology and thus deeply embedded in narratives of the modern.[16] The bourgeoisie understood in multiple ways the power that steam technology generated. It was simultaneously a utopian project, the perspective of a governing class, and a reflection of an emergent modernity—a folding together of grand narrative, ideological outlook, and raw history that is now difficult to pry apart. An explicitly class project, this (what Malm names) "steam fetishism" connected technology and science to progress, wealth, private property, and freedom—that is to say, to a liberal capitalism both propelled by and legitimated through technological innovation.[17] The language of solar transition depends on the selfsame set of equivalences (i.e., progress, freedom, wealth, and the rest), with one important difference. Those who commit to solar transition can claim to bring an ethical perspective to energy history. An emergent green bourgeois ideology now disavows steam as an energetic mistake due to the civilizational and environmental consequences of

15. As but one example, see Hermann Scheer, *Energy Autonomy: The Economic, Social and Technological Case for Renewable Energy* (New York: Routledge, 2018).

16. See Malm, *Fossil Capital,* 194–222.

17. For a variation of this argument, see Matt Huber, "Refined Politics: Petroleum Products, Neoliberalism, and the Ecology of Entrepreneurial Life," *Journal of American Studies* 46, no. 2 (2012): 295–312.

its use. But remarkably, doing so doesn't thus unnerve the original connections established in steam fetishism between technology and wealth, science and freedom, or steam and history. One might have expected the shifts and realignments of liberal ideology from dirty to clean energy to have unnerved the underlying fantasy of liberal capitalism, once the energy lie of modernity is revealed, as many critics of capitalism's long-term role in climate crisis expect it should be.[18] Instead, a commitment to solar transition manages the feat of transfiguring capitalism from the principal cause of eco-collapse into the one force that has both the infrastructural capacity *and* the ethical authority to do something about it—a reaffirmation of liberalism's capacity to unfold toward a greater good, figured this time in relation to the type of energy in use.

Let's call this new ideology *solar fetishism,* an emergent twenty-first-century bourgeois ideology that, like its steam progenitor, combines utopian project, the perspective of a governing class, and a common sense now in the process of being nervously composed and imposed. Threatening to drain this fetish of its spiritual and ethical energy is the reality that for the moment, solar energy just doesn't pay: profit largely remains tied to the continued use of fossil fuels. As luck would have it, there are creatures living among us—or, rather, looking down at us from the heights of the 0.001 percent—for whom money doesn't matter, especially when it comes to doing the environmentally right thing. Meta-entrepreneurs, those extra- and quasi-state actors who act when states seem unable to do so, have the power to make solar profitable and, in the process, show how contemporary capitalism operates with the best interests of the planet in mind. Their efforts help to keep the magic of the fetish alive.

18. See, e.g., David Schwartzman, "Solar Communism," *Science and Society* 60, no. 3 (1996): 307–31. For another variant of this argument, see Ulrich Brand and Markus Wissen, *Everyday Life and the Ecological Crisis of Capitalism,* trans. Zachary King (New York: Verso, 2021).

Which brings us back to Bill Gates, one of the (supposedly) good billionaires, a figure who has been described (by a critic!) as "unquestionably an ally to the climate movement"[19] and from whom we can understand just how green ideology and solar fetishism manage to do their work.

Solar Fetishism in Action: Green Premiums, Self-Interest, and Care

For more than two decades, Gates has mobilized his wealth toward achieving social ends. The Bill and Melinda Gates Foundation, established in 2000, has supported projects to eradicate disease in developing countries—most famously through its vaccine programs—and to ameliorate what it sees as flaws and limits in the current U.S. education system. With a current endowment of more than US$70 billion and total grant payments of US$77.6 billion since its inception, the Gates Foundation has the resources to make substantial inroads in both health and education.[20] The foundation has found greater success in its health program than in its educational endeavors, though both have been criticized for their framing of problems and approaches to solutions and especially for their failure to involve impacted communities in program development.[21]

19. Leah Stokes, "Bill Gates and the Problem with Climate Solutionism," *MIT Technology Review,* February 16, 2021, https://www.technologyreview.com/2021/02/16/1017832/gates-robinson-kolbert-review-climate-disaster-solutionism/.

20. See the Gates Foundation fact sheet at https://www.gatesfoundation.org/about/foundation-fact-sheet.

21. For recent criticisms, see Linsey McGoey, "How Bill Gates Makes the World Worse Off: Interview with Nathan J. Robinson," *Current Affairs,* July 29, 2022, https://www.currentaffairs.org/2022/07/how-bill-gates-makes-the-world-worse-off, and Valerie Strauss, "Um, Who Are Melinda and Bill Gates Trying to Kid?," *Washington Post,* April 16, 2019, https://www.washingtonpost.com/education/2019/04/16/um-who-are-melinda-bill-gates-trying-kid/. McGoey is author of *No Such Thing as a Free Gift: The Gates Foundation and the Price of Philanthropy* (New York: Verso, 2016).

Among the titans of the digital era, Gates stands out as one of the few who have mobilized their wealth to effect social goals.[22] In *New Prophets of Capitalism,* Nicole Aschcoff describes Gates as a "philanthrocapitalist," a figure who wants "to harness the forces of capitalism that made them fabulously wealthy to help out the rest of the planet."[23] A range of other organizations undertake philanthropic work on the basis of enormous endowments received from their capitalist progenitors, the most famous being the U.S. foundations established at the beginning of the twentieth century (e.g., Carnegie, Ford, and Rockefeller). Aschcoff draws a distinction between these foundations and some of those more recently established, especially the Gates Foundation. Whereas typical foundations "believe in old fashioned charity," philanthrocapitalists have larger ambitions: "[they] think profitable solutions to social problems are superior to unprofitable ones because they give private capital an incentive to care."[24] Philanthrocapitalists—a small but growing brand of meta-entrepreneur—approach their missions through the neoliberal rationality from which their wealth emerges.[25] They have become extraordinarily wealthy from a system that fuels their capacity to do so; this selfsame sys-

22. By comparison, the Bezos Earth Fund, established by ex–Amazon CEO Jeff Bezos, has distributed only US$1.4 billion. See Rachel Sandler, "With Jeff Bezos' Latest Grants, Here's How Far along He Is on His $10 Billion Earth Fund Giving," *Forbes,* December 6, 2021, https://www.forbes.com/sites/rachelsandler/2021/12/06/jeff-bezos-just-gave-433-million-to-climate-groups/.

23. Nicole Aschcoff, *The New Prophets of Capital* (New York: Verso, 2015), 108.

24. Aschcoff, 108.

25. A recent example of a plunge into philanthrocapitalism is the much publicized decision by Patagonia owner Yvon Chouinard to turn the company into a nonprofit foundation. See Matt Ott and Glenn Gamboa, "Patagonia Founder Gives Company Away to Environmental Trusts," Associated Press, September 15, 2022, https://apnews.com/article/patagonia-founder-donation-environment-55eaba4b90c9c6a271e75fc8d906545e.

tem, within which states have been drained of resources and (or so the rhetoric of neoliberalism asserts) which lacks the intellectual capacity and moral fortitude to effectively resolve problems, has necessitated that philanthrocapitalists step in for the broader social good, employing their business smarts, leadership experience, wealth, and unparalleled influence to effect change.

Gates has only recently turned his attention to saving the planet. In his best-selling 2021 book *How to Avoid a Climate Disaster: The Solutions We Have and the Breakthroughs We Need,* he names 2015 as the moment when he committed himself to saving the planet.[26] Gates points to two events that finally solidified his worries about energy and the environment—concerns that had lingered in the background in much of his work with the Gates Foundation. Each moment offers a synecdoche of how Gates views the specific contribution he believes he can make to climate action as a meta-entrepreneur. In the first case, he describes how he became aware of the student-led movement demanding university endowments divest from fossil fuels. A campaign launched by the *Guardian* asked the Gates Foundation to do the same: to sell its investments in the fossil fuel industry.[27] The foundation eventually did so, a decision Gates describes as a "personal choice."[28] Nonetheless, he chastises the students and the *Guardian* for what he sees as a fundamental misread of the politics of climate change, noting that divestment has little chance of effecting an actual reduction in CO_2 levels. He also suggests that divestment movements fail to attend to energy justice, arguing that he "didn't think it was fair for anyone to tell Indians that their children couldn't have lights to study by, or that thousands of Indians should die in heat waves because installing

26. Bill Gates, *How to Avoid a Climate Disaster: The Solutions We Have and the Breakthroughs We Need* (New York: Knopf, 2021).

27. "Keep It in the Ground" campaign, *Guardian,* https://www.theguardian.com/environment/ng-interactive/2015/mar/16/keep-it-in-the-ground-guardian-climate-change-campaign.

28. Gates, *How to Avoid a Climate Disaster,* 10.

air conditioners is bad for the environment."[29] The link Gates draws between energy use in developing countries and stocks in university endowments is, of course, a straw man argument. But the rhetorical effect of the argument is real enough. It allows Gates to evidence his mature read of a situation and concern for the experience of others, specifically the energy challenges those living in the Global South face. This concern for the developing world is foregrounded throughout his book (and on his website), which works effectively to diffuse criticisms that might otherwise be voiced about his positions on climate action. Gates evidences common sense; the students are painted as well-intentioned creatures who don't yet know the ways of the world.

The second event Gates recounts is equally revealing about his view as to what constitutes effective climate action. He describes what investment in new technologies and innovation can accomplish to reduce the production of greenhouse gases. Gates recounts steps he took in advance of the 2015 UN Climate Change Conference (more commonly known as COP 21) in Paris to mobilize governments and investors to generate funding for green tech, a sector in which returns had become low, thus leading investors to pull out their money. By the beginning of COP 21, Gates had brought together twenty-six investors and formed a new investment fund called Breakthrough Energy (members now include Jeff Bezos, Michael Bloomberg, Richard Branson, Jack Ma, George Soros, and Mark Zuckerberg).[30] He brought governments into the project of green energy as well, convincing French president François Hollande, U.S. president Barack Obama, and Indian prime minister Narendra Modi to double the green tech funding of their respective governments.

29. Gates, 9.
30. See https://breakthroughenergy.org/. Criticisms of the group's commitment to nuclear energy were raised almost immediately upon its creation. See Linda Pentz Gunter, "Is Gates's 'Breakthrough Energy Coalition' a Nuclear Spearhead?," *Ecologist,* December 6, 2015, https://theecologist.org/2015/dec/06/gatess-breakthrough-energy-coalition-nuclear-spearhead.

The difference between Gates's eco-accomplishment and that of the divestment students couldn't be clearer: in contrast to what he sees as the ineffective and misdirected protests of students, Gates shows that he has the ear of business and government and can make big things happen in a flash. The Paris Agreement signed during COP 21, which has become a touchstone in global efforts to address climate change, is never mentioned by Gates. And why should it be? The real action on climate took place elsewhere, before governments and nongovernmental organizations (NGOs) even showed up in Paris to do their work.

In many ways, *How to Avoid a Climate Disaster* is an unobjectionable book—which is not the same as saying there aren't many aspects of it that are worrying.[31] It belongs to an increasingly familiar genre of climate texts in which the narrative is driven by the author's stated desire to "cut through the noise" (Gates actually uses this phrase[32]) of climate change by describing where things are and what can be done to resolve the climate crisis we are experiencing.[33] One might draw on far worse texts to provide an account of climate change and climate action (Tom Rand's 2020 *The Case for Climate Capitalism: Economic Solutions for a Planet in Crisis* is an especially bad one).[34] The main point Gates wants

31. See, e.g., Million Belay, "Bill Gates Should Stop Telling Africans What Kind of Agriculture Africans Need," *Scientific American,* July 6, 2021, https://www.scientificamerican.com/article/bill-gates-should-stop-telling -africans-what-kind-of-agriculture-africans-need1/. The nonprofit investigative research group U.S. Right to Know keeps track of reports and news articles critical of the Gates Foundation's health and agricultural projects in Africa. "Critiques of Gates Foundation agricultural interventions in Africa" can be found at https://usrtk.org/bill-gates/critiques-of-gates-foundation/.

32. Gates, *How to Avoid a Climate Disaster,* 20.

33. See, e.g., Christiana Figueres and Tom Rivett-Carnac, *The Future We Choose: The Stubborn Optimist's Guide to the Climate Crisis* (New York: Vintage, 2021), and Paul Hawken, *Regeneration: Ending the Climate Crisis in One Generation* (New York: Penguin, 2021).

34. Tom Rand, *The Case for Climate Capitalism: Economic Solutions for a Planet in Crisis* (Toronto, Ont.: ECW Press, 2020).

to make is stated unambiguously and is in line with what many (perhaps even most!) environmental groups believe needs to be done. The world adds fifty-one billion tons of greenhouse gases to the atmosphere annually; it must get to zero tons as quickly as possible. In *How to Avoid a Climate Disaster,* Gates investigates the five main activities that produce greenhouse gases—making things, generating power, growing things, getting around, and keeping warm and cool.[35] In each case, he carefully enumerates emissions, outlines the effectiveness of technological solutions, and provides ideas for possible ways forward. The book avoids any number of possible missteps. No easy solutions are proffered (the title of the second chapter is "This Will Be Hard"). It is attentive to the distinct energy and environmental challenges and experiences of the Global South. And on more than one occasion, Gates offers a mea culpa about his own privilege ("It's true that my carbon footprint is absurdly high"[36]). He repeatedly rejects degrowth strategies as ill considered, misinformed, and unethical. He believes that to enable development in the Global South, it should be possible for everyone to use *more* energy, not less. The climate challenge comes down to making sure that every one of the five activities Gates names is done in a way that makes zero impact, while improving the quality of life of every person on the planet.

Getting to zero is a laudable goal. It is *how* Gates proposes to get to zero that is troubling, especially in the context of his power and desire as a philanthrocapitalist to effect change in some ways (e.g., mobilizing huge investments in technology) while expressing disdain or disinterest in others (e.g., supporting or encouraging divestment). *How to Avoid a Climate Disaster* is a book that consciously sidesteps the politics of climate change to focus on where Gates believes the real action on climate change takes place. He writes,

35. These five descriptors are used on the Breakthrough Energy website.

36. Gates, *How to Avoid a Climate Disaster,* 14.

In the United States especially, the conversation about climate change has been sidetracked by politics. Some days it can seem as if we have little hope of getting anything done.

I think more like an engineer than a political scientist, and I don't have a solution to the politics of climate change. Instead, what I hope to do is focus the conversation on what getting to zero requires: We need to channel the world's passion and its scientific IQ into deploying the clean energy solutions we have now, and inventing new ones, so we stop adding greenhouse gases to the atmosphere.[37]

Throughout the book, politics is viewed as the problem, while technological advances constitute the only true solution to climate change. At times giddily, Gates describes emergent technologies—often ones in which he or Breakthrough Energy have invested—and their prospects for getting greenhouse gases down to zero. This emphasis on technology comes as little surprise, and not just because of Gates's own predilections: energy fetishisms, whether of the steam or solar variety, depend on it.

But to see this book as primarily about technology is to miss the real force of its argument and its implications for an emerging common sense on climate action. *How to Avoid a Climate Disaster* is a book whose main concern is to explain how and why capital must play a fundamental role in creating climate solutions. The route to zero emissions passes through capital; technological innovation can happen only when investors have turned their heads toward solar transition and have made the decision to redirect financial resources from fossil fuels to solar energy. For Gates, it's easy to explain why this hasn't yet happened. "There's a very good reason why fossil fuels are everywhere," he writes. "They're so inexpensive. As in, *oil is cheaper than a soft drink*."[38] Fossil fuels, whether oil, gas, or coal, are cheap, and the ease which with they can be used in existing infrastructure, combined with their affordability, means there is still little incentive to adopt green energy on a mass scale and even

37. Gates, 14.
38. Gates, 39.

less incentive for investors to put their money into it. The challenge Gates sets himself is to understand how to produce the economic incentive to entice capital to throw its resources into the transition from dirty to clean energy.

The metric of capital incentive is captured in Gates's neologism "Green Premium," which he uses to measure the difference between the cost of fossil fuels and the forms of green(er) energy fuels that might act as substitutes. For instance, the Green Premium between the retail price per gallon of gasoline (US$2.43) and advanced biofuels (US$5.00) is 106 percent,[39] while the premium for cement is calculated as a range of 75 to 140 percent (US$125 average per ton at present, US$219–300 per ton using carbon capture).[40] It is certainly possible to challenge (in the examples offered) whether either advanced biofuels or carbon capture constitutes the best substitute for fossil fuels, or whether any of the energy substitutes Gates proposes are the right ones. But to limit one's attention to this point is to miss the forest for the trees (or perhaps a more appropriate idiom would be that it is to miss the planetary climate disaster by spending one's time complaining about the need for anyone to fly). What matters more in understanding how Gates shapes a solar fetishism appropriate to the era is tracking his proposed solutions for the Green Premiums that exist today.

Gates suggests two ways to address Green Premiums, that is, to cut them down so that the price of green energy becomes the same as or less than the price of fossil fuels. The first is simply to have consumers pay the difference. To this solution, however, he offers a quick rejoinder: the premiums would need "to be made so low that everyone"—that is, countries around the world—"will be able to decarbonize."[41] The care Gates takes to note the unequal impacts of energy costs on different communities is evidenced here again.

39. Gates, 144.
40. Gates, 107.
41. Gates, 61.

The second option is to somehow close the gap between the costs of oil and of green energy. Using one of the examples offered earlier, this would mean making advanced biofuels the same price as gasoline, so that airplanes can fly while drastically reducing their production of greenhouse gases. The gap can be closed in several ways. For Gates, the fundamental and perhaps only way is through developing new technologies. Green Premiums can also be made smaller or eliminated altogether by governments, which can produce policies to support the adoption of new technologies (e.g., "it's the only way nuclear will have a chance of helping with climate change") or, alternatively, provide investors and entrepreneurs with financial resources to offset the costs of research and development.[42] In either case, closing the gap means creating a situation in which capital is willing to invest in developing new green technologies because they believe they will get returns. To date, neither business nor government in concert with business has managed to fully close the gap on its own. This is where the philanthrocapitalist comes in—a figure with the capacity to push a stalled system toward a desired goal because she can bring business and government together to common ends, as evidenced by Gates's spur-of-the-moment creation of Breakthrough Energy in advance of COP 21. In this account, the unique capacity of the nation-state to act on climate change doesn't necessitate new forms of nationalism; it requires meta-entrepreneurs to take the reins of power, acting for everyone's good.

In a speech he delivered at the 2008 World Economic Forum, Gates outlined what we might take as his base understanding of the fundamental operations of sociality:

> As I see it, there are two great forces of human nature: self-interest, and caring for others. Capitalism harnesses self-interest in a helpful and sustainable way, but only on behalf of those who can pay. Government aid and philanthropy channel our caring for those who

42. Gates, 191.

can't pay. But to provide rapid improvement for the poor we need a system that draws in innovators and businesses in a far better way than we do today.[43]

These words were articulated in reference to the Gates Foundation's support for global health initiatives; in 2008, climate change was not an issue that had yet captured Gates's full attention. But it is easy to see how the divide between self-interest and care explicated here animates his views on climate action as well. To start to grasp how he imagines the role philanthropists—and specifically philanthrocapitalists—can play in addressing climate change, and to understand, too, the specific function of capitalism in climate action, all one need do is substitute "environment" for "the poor" in the preceding statement.

And what does this substitution tell us about climate action? It is that care plays an essential role in producing the conditions that will allow the gap represented by Green Premiums to be closed; only care—the opposite of self-interest—can initiate a process through which the power of markets to make rapid improvements can be mobilized in support of issues that fall outside of their episteme, such as the poor or the environment (neither of which can pay for anything they may need). The inability of capitalism to attend to things it should care about should constitute a substantial rebuke to its capacity to act on climate or, indeed, to deal with anything other than that which can be rendered profitable. But Gates sees things differently. He suggests that capitalism has the capacity to harness "self-interest in a helpful and sustainable way." Gates believes that when it comes to addressing huge problems like poverty and climate change, it would be a mistake to sidestep capitalism's capacity to generate unique solutions. In his view, the function of care isn't to make the poor and the environment safe *from* capitalism but to make them safe *for* capitalism to do the self-interested work

43. Bill Gates, "A New Approach to Capitalism in the 21st Century," speech at the World Economic Forum 2008, Davos, Switzerland, January 24, 2008.

it does best. No class is better equipped to do just this than philan-throcapitalists. Only Gates and others like him have the power to bridge the divide between self-interest and care, which is necessary to save us from disaster.

How to Avoid a Climate Disaster captures the form and charac-teristics of an emerging worldview on energy and environmen-tal futures. It is a worldview that accords perfectly with existing neoliberal understandings of how the state and capital can most effectively operate together to resolve problems. Gates positions the state and the social—what was once the realm of care—on the outside looking in on the design space of technological problem solving. He concludes his book by cheerfully narrating the need for resources to flow from the public purse to the pockets of pri-vate investors, who will in turn sell their new technologies back to states in need of solutions. Care might be a way of characterizing the political motive guiding those activists, NGOs, governments, and international organizations that have worked tirelessly to ame-liorate the world created by self-interest. But if care is what really matters in bringing about environmental change, the proposition is that philanthrocapitalists can care better and in the right way, in concourse with self-interest. Philanthrocapitalists can act without the need for approval, debate, or dialogue and without a mandate secured by democratic governance. At a moment of political dys-function and climate crisis, Bill Gates's professed ability to put cap-italism to work for the environment and to suture self-interest and care positions him as a climate leader who, like a latter-day Moses, mobile phone raised overhead to get a signal in the desert, can lead us to the promised land of zero carbon.

Philanthrocapitalists and Common Sense

The worrisome and dangerous shift of power, wealth, and decision-making from public to private has been examined in numerous stud-ies of the logics and practices of neoliberalism. The brute realities of climate change, now directly felt by rich and poor (though in far different ways) through extreme weather events, have been seen by some as creating an opening for a critique of capitalism as a system with disastrous ends. But what happens if capitalists themselves begin to publicly recognize the civilizational crisis capitalism has induced? What happens to the effectiveness of environmental critique if capitalists begin to commit resources to solving the problems their class has generated? Through his actions and the work of his foundation, Gates the philanthrocapitalist argues that capitalism can be caring, if only given the right incentives. The concerted pushback on energy transition by some capitalists—oil executives and the like—only further reinforces the social and environmental concerns endemic to newer, shinier, and greener forms of techno-capitalism and the capacity of good meta-entrepreneurs to solve meta-problems for the good of all.

Despite its many contradictions and inconsistencies (capitalists addressing the consequences of capitalism . . . through capitalism?), there is an undoubted power to eco-ideologies shaped by meta-entrepreneurs. The articulation of solar fetishism further solidifies already prevalent beliefs in technology as the primary driver of history—the solar panel lying just a few pages ahead of the steam engine in the textbook of liberal capitalism, helping the whole narrative to remain coherent even in the face of the potential end of humanity. And the actions of meta-entrepreneurs defer the need for publics to act. In the face of a complex problem like climate change, citizens are gifted the opportunity to stop worrying: the meta-entrepreneur has the unique formula to figure out how to push stuck systems back into operation without the need to upend sociopolitical or economic apple carts in the process. How could anyone resist such a narrative? "Three Cheers for the Dull, Factually Correct Middle" is the title Gates gives to his review of Vaclav Smil's *How the World Really Works.*[44] There is no doubt about how he

44. See Gates on Smil, in a review published on the former's website: https://www.gatesnotes.com/Books/How-the-World-Really-Works.

envisions the proposals he makes in his own book—as politically inconsequential and without harm. It is in this dull middle of correct facts where incentive to change disappears into the givenness of who can best lead us to a greener world. And it is in this middle defined by facts and figures, and thus supposedly free of all politics and ideology, that we see the construction of a common sense of eco-capitalist action.

Gates's narrative of our energy and environmental futures is a vision shared by many others within industry who recognize the reality of climate change and want to act on it. As might be expected, this narrative is not without challengers. Those mounted against Gates and against the legitimacy of nation-states as climate actors are the subject of lecture 3, which concerns right-wing populisms and the answer they propose to the question of who owns the sun.

Lecture 3. From Convoys to Commons

When Snow Refuses to Melt

One might come away from my discussion of Bill Gates and his meta-entrepreneurial comrades with the impression that they are figures who are universally loved. And why wouldn't they be? They seem able to do things. While governments hold back on making important decisions, about the environment and other matters too, entrepreneurs make things happen. With their space programs, Elon Musk and Jeff Bezos have managed to do what once only governments could. And the small modular reactors (SMUs) championed by Bill Gates and Breakthrough Energy stand in sharp contrast to the slow retrofits and even slower build of new nuclear facilities by public utilities—that is, if the latter are even willing to take up the task. That SMUs don't yet truly exist doesn't matter; it's the *perception* of action by meta-entrepreneurs in contrast to governments that renders the latter less (and less) important to shaping the direction the future will take—not entirely unimportant, just less so than in the past.

The love for Gates only runs so deep. His apparent power and his comfort (and self-satisfaction) in using this power to shape the direction of energy transition and climate change have been criticized as antidemocratic and environmentally dangerous. For example, his strong faith in the use of industrial fertilizers and monoculture

crops to feed the world has been challenged by advocates of regen-
erative agriculture, including Vandana Shiva. Bill McKibben, has
described Gates's proposals for energy transition as "surprisingly
behind the curve on the geeky parts, and . . . worse at interpret-
ing the deeper and more critical aspects of the global warming
dilemma."[1] But these are not even the most severe or significant
criticisms that have been made of Gates. In some communities, he
is openly despised. The use of his name is commonplace in widely
circulated right-wing conspiracy theories proposing to identify the
cabal of figures in control of the direction in which we're all heading.

It's difficult to identify precisely when Gates first became a synecdoche in conspiracy theories for misguided, dangerous, and (supposedly) left-wing, technocratic power in general. A search of the
phrase "Bill Gates conspiracy" on Google Trends shows that the
connection between these terms extends back to at least 2004, with
discrete spikes in the subsequent two decades. The highest spike
occurred in April 2020, during the initial months of the Covid-19
crisis, when Gates became looped into conspiracy narratives about
the origin of the SARS-CoV-2 virus. His presumed role in the creation of the virus, and the development of a vaccine to deal with
it, has for some elements of the right offered the clearest evidence
yet that there is a nefarious scheme by meta-entrepreneurs to rule
the world.

In a 2015 TED talk available on YouTube, Gates spoke of the need
to prepare a response to a virus that might imperil life around the
world, just as the Ebola virus had threatened to do in 2014, when
the WHO declared an outbreak in West Africa to be an internation-

1. Vandana Shiva, "Bill Gates Empires 'Must Be Dismantled':
Interview with Dr. Joseph Mercola," *Regeneration International,* April 5,
2021, https://regenerationinternational.org/2021/04/05/vandana-shiva-bill
-gates-empires-must-be-dismantled; Bill McKibben, "How Does Bill Gates
Plan to Solve the Climate Crisis?," *New York Times,* March 9, 2021, https://
www.nytimes.com/2021/02/15/books/review/bill-gates-how-to-avoid-a
-climate-disaster.html.

al health emergency.[2] This video, at time of writing having been viewed by more than thirty-seven million people, was stumbled upon by anti-vaxxers, QAnon, and other right-wing pundits, including Alex Jones, Robert F. Kennedy Jr., and *Fox News* host Laura Ingraham, and transformed into a powerful political meme. Gates's video made it possible to reimagine his sharp, public criticism of the failure of the Trump administration to mount an adequate response to Covid-19 as clear evidence of a complex plot. Descriptions varied as to just what the plot was: Gates had created SARS-CoV-2 to profit from vaccines created to address Covid-19; he was using the illness to kill off segments of the population; or he had implanted a chip into the vaccine that would allow him to control the planet's population—a proposition that beautifully conjoined his work with Microsoft and the activities of the Gates Foundation. In its report on the conspiracy theory, the *New York Times* wrote that Gates had quickly "assumed the role occupied by George Soros, the billionaire financier and Democratic donor who has been a villain for the right."[3] The ill-timed decision in July 2021 by a consortium led by Gates and Soros to purchase Covid-19 test manufacturer Mologic further hinted at something afoot behind the curtains of power and pushed the Gates conspiracy theory in a new direction.[4] In January 2022, Michael Flynn, former three-star general and a leading figure in the QAnon movement, claimed that SARS-CoV-2 had been created by Gates, Soros, and others—Klaus Schwab, chairperson of

2. Bill Gates, "The Next Outbreak? We're Not Ready," YouTube video, April 3, 2015, 8:36, https://www.youtube.com/watch?v=6Af6b_wyiwI.

3. The most best single account of the Gates conspiracy theory is offered by Daisuke Wakabayashi, Davey Alba, and Marc Tracy, "Bill Gates, at Odds with Trump on Virus, Becomes a Right-Wing Target," *New York Times,* April 17, 2020, https://www.nytimes.com/2020/04/17/technology/bill-gates-virus-conspiracy-theories.html.

4. David Dawkins, "George Soros and Bill Gates' Backed Consortium to Buy U.K. Maker of COVID Tests for $41 Million," *Forbes,* July 19, 2021, https://www.forbes.com/sites/daviddawkins/2021/07/19/george-soros-and-bill-gates-backed-consortium-to-buy-uk-maker-of-COVID-lateral-flow-tests-for-41-million/.

the World Economic Council among them—to steal the 2020 U.S. election.[5] It's a claim that has to be admired for its elegance: Flynn managed to pack all the current complaints of the U.S. right into a single conspiratorial assertion.

It doesn't really need to be said that these claims are inaccurate—or, let's be frank, unhinged. Covid-19 vaccines do not contain microchips for the purposes of surveillance. Cell phones, CCTV cameras, and social media companies do this work well enough; a meta-entrepreneur who lives and breathes the air of cost–benefit analyses would never waste time and money on something so costly and complex as a vaccine chip. What these narratives confirm, however, is a belief in the enormous power of individuals like Gates to determine the shape of reality and to do so without the need for either government input or approval from members of the general public. In his account of the Gates–Soros conspiracy, Flynn reminds his listeners on Twitter that the duo has "placed themselves above some . . . world institutions and they drive them, they really do drive them."[6] It would be a mistake to see these conspiracies as simply antibusiness or as opposed to the power of all meta-entrepreneurs. Peter Thiel, former board member of Meta and one of the largest donors to the U.S. Republican Party, hasn't been the subject of conspiracy theories, nor has (for complicated reasons) Elon Musk, despite his sketchy proposals to end the war in Ukraine and single-handedly resolve the conflict over Taiwan—world-historical acts that only a meta-entrepreneur would imagine himself being able to carry out.[7] What differentiates Gates from ultrarich figures like

5. Graeme Massie, "Michael Flynn Claims Globalists Will Try to Turn Humans into Cyborgs after Failing to Change Their DNA with COVID Shots," *Independent*, January 21, 2002, https://www.independent.co.uk /news/world/americas/us-politics/COVID-19-vaccine-michael-flynn-dna -b2172413.html.

6. Flynn's video is available at https://cdn.jwplayer.com/previews /CZQZfETl.

7. Ryan Mac and Mike Issac, "Peter Thiel to Exit Meta's Board to Support Trump-Aligned Candidates," *New York Times*, February 7, 2022,

Thiel or Musk is not only the political party he supports but also that Gates has made it his explicit task to change the world for the better, even if—as I argued in my last lecture—this turns out to be a change that pretty much keeps everything the same as it is now.

The Gates conspiracy theory isn't limited to his supposed role in Covid-19. Winter Storm Uri, which in 2021 caused millions of homes in Texas to lose power, generated theories about the role Gates may have played in making it happen. Instead of recognizing the storm as a climate change–induced weather event, some saw it as a geoengineering experiment—an attempt to block the sun to cool down the earth, even at the price of leaving millions in the cold. A series of TikTok videos attempted to confirm this by showing users trying to melt snow left by Uri with matches or hair dryers, only to find that the ice refused to turn to water.[8] Drawing a link between Gates and geoengineering isn't purely science fiction: a research project at Harvard University called SCoPEx (for Stratospheric Controlled Perturbation Experiment) has received funding from Gates, and although Gates points to some problems with geoengineering in *How to Avoid a Climate Disaster,* he doesn't rule it out as a potential solution to end climate change. Indeed, he writes, "geoengineering is the only known way that we could hope to lower the earth's temperature within years or even decades without crippling the economy. There may come a day when we don't have

---

https://www.nytimes.com/2022/02/07/technology/peter-thiel-facebook
.html; Rita Trichur, "Elon Musk Has Gone Too Far. It's Time to Boycott Tesla," *Globe and Mail,* October 14, 2022, https://www.theglobeandmail.com
/business/commentary/article-elon-musk-tesla-boycott/. The Musk poll on Ukraine can be found on X (formerly Twitter) at https://x.com/elonmusk
/status/1576969255031296000.

    8. Rachel E. Greenspan, "TikTokers Are Trying to Prove That Snow in Texas Is 'Fake,' Pushing a False Conspiracy Story," *Insider,* February 22, 2021, https://www.insider.com/fake-texas-snow-not-melting-tiktok
-conspiracy-theory-2021-2. The hashtag #governmentsnow was also being used on TikTok at the time to explain the cause of Texas snow; that is, the Biden administration was also in on the Gates conspiracy.

a choice. Best to prepare for that day now."[9] It may seem perverse to suggest that the conspiracy minded are correct in their view of Gates. But the feeling of many that others are making decisions about their lives is real enough, no matter if attempts to make sense of this can be confusing, hypocritical, contradictory, delusional, or even dangerous.

Recent right-wing challenges to the legitimacy of the government and science making decisions on the public's behalf have both unnerved institutions and limited their capacity to address climate change and other social issues. For this reason alone, an all-too-common response to the kinds of conspiracies I've been describing, and to the right-wing attempts to animate and amplify them, has been to quickly reaffirm the status quo role of government and business in political life. After all, who else can be counted on to show publics that there might be reasons why it might snow in Texas that don't require secretive lever pulling? But the production of such a counterposition, which seems so obvious and essential, contains ideological dangers of its own. The political threat of the zany *Struwwelpeter* narratives that make up right-wing conspiracy theories doesn't lie only in the distractions and confusions it imposes on some segments of the populace; such narratives can't help but make it seem that the better—or, indeed, the *only*—option might be Gates or what I have described in my first lecture as "good nationalism," each of which is animated by its own fantasy of its unique capacity to take on climate change in the most efficacious way possible. Conspiracies can quickly transform meta-entrepreneurial power and fictions of good liberal nationalism from propositions into a common sense about which the rest of us can agree, the illegitimacy of the former making the latter two narratives all the more convincing.

As I suggested at the outset of these lectures, there is an ongoing struggle by several actors to shape a common sense around energy

9. Gates, *How to Avoid A Climate Disaster*, 178.

transition and action on climate change—a common sense that can be relied upon (its proponents assert) always to take appropriate and reasonable climate action, if and when any is deemed necessary. The globalist eco-pretensions of the contemporary liberal capitalist state constitute one claim on common sense; another can be found in the "ethical" neoliberal entrepreneurialism that Gates embodies so well. A third is to be found, perhaps counterintuitively, in practices—real and virtual—of an ideology and movement *opposed* to energy transition and climate action: the mess and muddle of contemporary right-wing reactions to all matters concerning the environment across the globe. This opposition takes many forms, from contemporary eco-fascism, which conjoins present-day environmental discourses about protecting the earth and violent, *Blut und Boden* nationalisms, to the views held by some members of the religious right who see no contradiction between God's assignation to man of dominion over the earth and species loss and strip-mining.[10] In the other two cases I have examined, the radical action required to address climate change is tempered by the desire to keep political and economic systems much the same as they are at present. The mess of right-wing anti-eco positions works to the same end, by affirming the necessity of contemporary extractive practices in the name of tradition, thus shaping a reactionary environmental conservatism to match the sociopolitical ones it advocates. For understandable reasons, much discussion of contemporary right-wing politics has focused on the United States. But what I now want to examine is the right-wing extractive politics of its neighbor Canada, where the right to use fossil fuels has been sutured together with libertarian freedoms via conspiracy theories, ideological projects launched by governments committed to Big Oil, and nationalist

10. On eco-fascism, see Casey Williams, "Fossil Fuels, Climate Change, and the Modern Crisis of Imagination" (PhD diss., Duke University, 2022). On religion and environmentalism, see Darren Fleet, "Fuel and Faith: a spiritual geography of fossil fuels in Western Canada" (PhD diss., Simon Fraser University, 2021).

narratives circulated by the fossil fuel industry itself. What better way to examine how this third common sense is being constituted than by looking closely at a country in which so many elements of society seem to believe that fossil fuels matter more now that they are beginning to matter less?

"Pierre Elliot Trudeau Rips Off CANADA"

References to Bill Gates as master conspirator now have a useful analytic function: the appearance of his name in articles and documents, on websites, and in other forms of social media can be easily used to pinpoint the ideological leanings of their authors. For this reason alone, it was not entirely surprising to find Gates noted in Dr. T. L. Nemeth's April 2020 report *A New Global Paradigm: Understanding the Transnational Progressive Movement, the Energy Transition and the Great Transformation Strangling Alberta's Petroleum.*[11] This report was one of a number commissioned by the Public Inquiry into Anti-Alberta Energy Campaigns, launched by the Government of Alberta to investigate the role foreign funding played in campaigns waged by NGOs, research institutes, and other organizations intent on ending extraction in the oil sands. The first few sentences of Nemeth's report provide an immediate sense of her understanding of the situation. She writes, "There is a transnational global movement to facilitate a fundamental paradigm shift, a Great Transformation, to a new energy economy that will halt fossil fuel use and development, initially in the western world, in order to create a new global low-carbon, net-zero civilization."[12] This opening sentence is unobjectionable—there *is* just such a movement to create a net-zero world powered by clean energy and greener

11. T. L. Nemeth's report *A New Global Paradigm: Understanding the Transnational Progressive Movement, the Energy Transition and the Great Transformation Strangling Alberta's Petroleum,* April 2020, can be found at https://blog.friendsofscience.org/wp-content/uploads/2021/02/Nemeth -Report.pdf.

12. Nemeth, 3.

economic practices. But Nemeth's second sentence makes clear that her take on this movement is far from usual. She proclaims, "The first major petroleum industry target being used to accelerate this global transition is the Alberta oil sands." For Nemeth, as for the Alberta government, the great transformation of the energy economy constitutes a problem instead of a potential *solution.* Nemeth's 133-page report—one of a group of three for which the government paid Can$100,000—lays out in detail her understanding of how this problem came to be.[13]

Nemeth names Gates and his coconspirators as among the most significant actors in the challenges being mounted against extraction in the oil sands and the use of fossil fuels more generally. His coconspirators include (here we go again . . .) Klaus Schwab and George Soros, who are identified as having produced the "coronavirus pandemic in order to anticipate responses and gaps in preparedness" in right-wing challenges that might be mounted against unwanted transformations in the current energy system.[14] Nemeth finds that the cabal extends well beyond Gates et al. to include all those assembled at Davos in 2020, where, she notes, "climate change was among the seven main themes"—a fact that constitutes her main point of evidence.[15] The list of organizations Nemeth identifies as involved in the attack on Alberta's extractive economy is too long to reproduce in full; among them are Deloitte, Hitachi, Huawei, Kaiser Permanente, McKinsey, Microsoft, Mitsubishi Chemical, Saudi Aramco, Visa, the American Heart Association, Netflix, and Johnson & Johnson—and, of course, Greta Thunberg and Prince (now King) Charles. (An intriguing cast of characters, to be sure.) Her report attracted derision from the mainstream Canadian press—so much

13. Lisa Johnson, "Alberta Inquiry Responses to Criticism It Commissioned Reports Based on Junk Science," *Edmonton Journal,* January 15, 2021, https://edmontonjournal.com/news/politics/alberta-inquiry -responds-to-criticism-it-commissioned-reports-based-on-junk-science. Nemeth received Can$27,840 for her work.

14. Nemeth, *A New Global Paradigm,* 50.

15. Nemeth, 50.

so that she publicly retracted some of her more extreme claims. By contrast, the Government of Alberta stood by it, and the report was one of the group of studies that commissioner Steven Allan had to consider in writing his final report about the Machiavellian conspiracies working to close shop on bitumen extraction. Allan found no evidence that activists had done anything wrong.[16]

The Public Inquiry into Anti-Alberta Energy Campaigns, which was tasked with the mission to fight misinformation about oil and gas, was one of the first initiatives of the Can\$30 million Canadian Energy Centre, established in 2019 by the new Conservative government of Jason Kenney. It was just the latest in a history of campaigns mounted by the Alberta government to challenge threats to what it sees as not just its most important industry but also a defining aspect of its political self-identity (not unlike Texas in the United States). Though modeled on U.S. government inquiries into Russian attempts to subvert the 2016 election, it was clear from the outset that the message the public inquiry wanted to send wasn't to foreigners but to Albertans, to bring them ever more strongly into the Conservative fold, and to the Canadian federal government, to tell them to back off.[17] The appearance of the Gates–Soros conspiracy

16. Lisa Johnson, "No Evidence of Wrongdoing Found in Allan Inquiry Report into 'Anti-Alberta' Campaigns," *Edmonton Journal,* October 21, 2021, https://edmontonjournal.com/news/politics/long-awaited-allan-inquiry -report-into-anti-alberta-campaigns-released; Kelly Cryderman, "Alberta Energy Inquiry Says No Wrongdoing By Anti-Oil-Sands Activists," *Globe and Mail,* October 21, 2021, https://www.theglobeandmail.com/canada/alberta /article-alberta-energy-inquiry-says-no-wrongdoing-by-anti-oilsands -activists/. The final report, which found no evidence that activists had done anything wrong, can be found at https://open.alberta.ca/dataset/public -inquiry-into-anti-alberta-energy-campaigns-report/resource/a814cae3-8dd2 -4c9c-baf1-cf9cd364d2cb.

17. One of the principal documents named by Steve Allan as a guide for the inquiry was the U.S. Senate minority report titled *The Chain of Environmental Command: How a Club of Billionaires and Their Foundations Control the Environmental Movement and Obama's EPA,* https://www .epw.senate.gov/public/index.cfm/2014/7/post-53280dcb-9f2c-2e3a-7092 -10cf6d8d08df.

theory in Nemeth's report indicates a shift of approach in a long-running federal–provincial dispute. Nemeth reframes a conspiracy born in the United States in relation to political power as being fundamentally about extraction, and she expands the list of conspirators to include all governments and businesses involved in developing a green economy that—were it to become a reality—would threaten Alberta's right to extract. Given its multiple aims and ambitions, it shouldn't be surprising that the message the current Alberta government is sending via the efforts of the Canadian Energy Centre is so convoluted: trust government (Alberta), but don't trust government (everywhere else, but especially the Canadian federal government); trust business (fossil fuel extractors), but don't trust business (Gates and industry competitors like Saudi Aramco); and trust the reports of a Canadian Energy Centre that, by virtue of its moniker, claims to work on behalf of all Canadians but is in truth an advocate for a single province.

The shape of the contemporary politics of fossil fuels in Canada is the outcome of an extended history of intergovernmental conflict over extraction that has now indelibly sutured the identity of the Canadian right to Big Oil. The origin of the enmity between Alberta and Canada extends back to the 1930s and the birth of the now defunct ultraconservative Social Credit Party. But the start of the disputes whose end result is the Canadian Energy Centre dates to the National Energy Program (NEP), which was initiated in 1980 by the Canadian federal government to (effectively) nationalize aspects of the industry; this was done in part to respond to the financial challenges of the 1973 oil crisis.[18] There was fierce and unrelenting opposition to the NEP in Alberta and Western Canada, led by gov-

18. The NEP was crafted around three main principles: "(1) security of supply and ultimate independence from the world market, (2) opportunity for all Canadians to participate in the energy industry, particularly oil and gas, and to share in the benefits of its expansion, and (3) fairness, with a pricing and revenue-sharing regime which recognizes the needs and rights of all Canadians." Government of Canada, *Budget 1980* (Ottawa: Department of Finance, October 28, 1980).

ernment and the oil industry, which helped to bring about its end in 1985. A popular bumper sticker at the time read "Let the Eastern bastards freeze in the dark"—a sentiment that remains in place up to the present and is kept alive by industry and government in Alberta, either working separately or—as is more common—working together. The perceived overreach of the federal government on resource control and management was also the impetus behind the formation of the far-right Canadian Alliance and Reform Parties, both committed to representing the interests of Western Canadians federally; these parties now form the heart of the contemporary Conservative Party in Canada and the United Conservative Party in Alberta.[19] To claim that the desire to control resources was the origin of right-wing populism in Canada would be an overstatement. But to claim that there isn't a deep interrelation between the two would be to miss why it is so important for the right in Canada to affirm resource extraction as an important element of Alberta's heritage and tradition, which it perpetually must defend from all those who might seek to end it. (One of the many things lost in the shuffle of this history is that Petro-Canada—a Canadian crown corporation that emerged out of the NEP—was one of the early players in the oil sands and played a key role in the development of the oil sands and the offshore Hibernia oil project.)

The threat posed by the NEP has had repercussions in the fossil fuel industry as well; in its wake, industry has ever more aggressively sought to shape public opinion on the import of extraction to life in Canada. I began these lectures by stating that the era of oil is over. There's no doubt that Canadian industry recognizes this, especial-

19. See Dean Bennett, "'Most Discriminated Against Group': Alberta Premier Pledges to Protect Unvaccinated," *Global News,* October 11, 2022, https://globalnews.ca/news/9189811/danielle-smith-sworn-in-albertas -19th-premier/; Max Fawcett, "Danielle Smith's Anti-expert Crusade Will Crash Alberta's Health-Care System," *National Observer,* October 24, 2022, https://www.nationalobserver.com/2022/10/24/opinion/danielle-smith -anti-expert-crusade-crash-alberta-health-care-system?.

ly because its main site of extraction—the Athabasca oil sands—has become a globally recognized sign of everything that is wrong with extractive economies. In response, the Alberta oil industry has sought to legitimize oil sands extraction in whatever way it can. It has supplemented the work sympathetic governments have done on behalf of industry, as exemplified by initiatives like the Alberta public inquiry, with an increasingly savvy undercover campaign intended to gnaw away at public opposition to its practices. To give one example of this, the recently established Pathways Alliance has publicly articulated a plan for how to achieve net-zero greenhouse gas emissions from the oil sands by 2050.[20] The Alliance was established by the six companies that together operate 95 percent of oil sands production: Canadian Natural, Cenovus, ConocoPhillips, Imperial, MEG Energy, and Suncor (the company that, in 2009, absorbed the by then private company Petro-Canada). A visit to the Alliance's website shows how expertly industry adapts the language of the Paris Agreement to its explanation of how it intends to achieve net zero. The now standard language of environmental goals to be achieved and newfound commitments to be made to sociopolitical issues (e.g., a land acknowledgment appears at the bottom on every page of the website) almost buries the lede: industry hopes to get to net zero not by reducing or eliminating extraction but by using new technologies, such as carbon capture and storage, to manage its greenhouse gas production.

The supposed commitment of the fossil fuel industry to the tenets of the Paris Agreement is just one approach to get Canadians to remain committed to extraction. Another has been to insist on the specific social and political importance of Canadian oil. The campaign launched by industry and government to explain that oil from the oil sands matters because of its *Canadianness* had two stages. The first was captured in Ezra Levant's 2010 *Ethical Oil: The Case*

20. https://pathwaysalliance.ca/.

for Canada's Oil Sands.[21] Levant's argument is simple. Fossil fuels are still necessary, and as most countries don't produce enough of their own, they must buy the stuff from somewhere. Most fossil fuels are produced in petro-states with poor human rights records and led by totalitarian governments unable to escape the "oil curse."[22] By contrast, Canada produces a uniquely "ethical oil," because it originates in a First World, democratic country known for being a global leader in human rights. In opposition to environmentalists who want to limit or stop oil sands production, Levant writes, "The question is not whether we should use oil sands oil instead of some perfect fantasy fuel that hasn't been invented yet. Until that miracle fuel is invented, the question is whether we should use oil from the oil sands or oil from the other places in the world that pump it."[23] In the years immediately following the publication of Levant's book, ethical oil was widely employed by industry to reinforce the legitimacy of continued extraction, until the term faded from widespread use. The notion of "ethical oil" was revived in 2018 by Ontario Proud, a right-wing group committed to returning "power to the people," in a series of ads and via its website StopSaudiOil.com—an older message perfectly suited to the xenophobic Trump era.[24]

A variant of this approach has been industry's frequent assertions that it represents Canadian values and plays an essential role in the life of the country. The Canadian Association of Petroleum Producers (CAPP) has mounted several increasingly sophisticated

21. Ezra Levant, *Ethical Oil: The Case for Canada's Oil Sands* (Toronto, Ont.: McClelland and Stewart, 2010). For an extended discussion of Levant, see my "How to Know about Oil: Energy Epistemologies and Political Futures," *Journal of Canadian Studies/Revue d'*études *canadiennes* 47, no. 3 (2013): 145–68.

22. On the notion of an "oil curse," see Michael L. Ross, *The Oil Curse: How Petroleum Wealth Shapes the Development of Nations* (Princeton, N.J.: Princeton University Press, 2013).

23. Levant, *Ethical Oil,* 6–7.

24. See Mike Ekers, "'Self' or 'Other'? A Multimodal Critical Discourse Analysis on Canadian Petro-nationalism Based on Social Media Posters," *Social Semiotics* (forthcoming).

campaigns along these lines. Its "Raise Your Hand for Canada" campaign shows pictures of (mostly white) Canadians holding a hand up, fingers spread, over which is pasted the outline of a maple leaf. The concluding lines of the ad copy ask the question, "Think energy developed the Canadian way is good for Canada?" The answer the CAPP wants to hear: "Then now is the time to say so by raising your hand." The latest campaign to marry oil with Canadian values was initiated in the wake of the war in Ukraine. The website "Made the Canadian Way" asks, "If your oil and gas came with a label, what would you learn?" The label posted online tells us that Canada's oil and gas are "innovation driven" and "responsibly produced." This campaign was mounted by the Canadian Energy Centre—which, if you remember, was created by the Government of Alberta to support industry in whatever way possible.

What is the takeaway from these government and industry campaigns? Are we supposed to understand fossil fuels as a good belonging to all Canadians, as industry ad campaigns seem to suggest, or to Albertans alone, which is the standpoint of the Alberta government? Is industry genuinely committed to decarbonizing, or do all such claims constitute little more than greenwashing?[25] The NEP is long dead. How, then, are we meant to understand the current relationship between the federal and provincial governments? Large segments of the public, both inside and outside of Alberta, continue to believe that the federal government is impeding the financial success of the fossil fuel industry and so must be resisted. While the federal government has begun to implement a carbon tax—the proceeds of which it returns to Canadians—it also continues to provide enormous subsidies to industry; during the pandemic alone, oil and gas industries received more than Can$18 billion in subsidies.[26] Four

25. On greenwashing in the energy industry, see Jordan Kinder, *Petroturfing: Refining Canadian Oil in the Age of Social Media* (Minneapolis: University of Minnesota Press, 2024).

26. Sarah Cox, "Canada's Oil and Gas Sector Received $18 Billion in Subsidies, Public Financing during Pandemic: Report," *Narwahl,* April 15, 2021, https://thenarwhal.ca/canada-oil-gas-pandemic-subsidies-report/.

decades on, the legacy of the NEP has been rewritten by successive right-wing governments and their industry comrades in a way that has made the private appear to be public (i.e., that the Canadian public owns oil, when it is in fact owned by private industry), even while managing to retain the notion that there is a danger that the federal government will nationalize the oil industry, that is, make public what ideologically has *already been* rendered public in most Canadian minds. Rendering oil ownership this way has proved to be incredibly effective common sense that has made any thought of changes to the dominant energy system in Canada hard to parse.

The political common sense I've just named is filled with contradictions—deliberately so—organized around oppositions that can flip-flop as needed, from being in favor of business to apparently being against it, from opposing nationalism to embracing it. What is certain, however, is that energy transition and environmental policies, even to a limited degree, have been positioned as threats to established political systems whose ongoing legitimacy deeply depends on continuing fossil fuel extraction. To the frustrated middle classes created by neoliberalism, industry and their governmental representatives have repeatedly offered an answer about what has gone wrong: misguided technocrats are messing with resources essential to your self-identity and livelihood, and they are doing so by employing globalist environmental policies that fail to attend to national identity (sometimes expressed in provincial guise). The result is that in Canada, the xenophobia of right-wing populisms has been shaped not only by exclusions defined by race, ethnicity, and sexual orientation but also by differences with respect to *energy resources*—a unique political configuration with serious consequences for energy transition and climate action in the country.

"Nothing Ever Happens in Ottawa"

Given the confusions I've just described, which shape belonging, tradition, private, and public into complex figures, it is perhaps unsurprising that adding one more ingredient to this stew of ideas, positions, and counterpositions would generate new uncertainties that could result in a crisis. The monthlong Freedom Convoy (also known as the Trucker's Convoy), which made its way across Canada before reaching Ottawa on January 29, 2021, was kicked off by just such an event, which mobilized the center–periphery narratives developed around extraction to new ends.[27] The extended occupation of downtown Ottawa in and around Parliament Hill was the immediate result of the imposition of vaccine mandates on truckers crossing the U.S.–Canadian border, which frustrated truckers who had decided not to be vaccinated. To show their displeasure at the vaccine mandate and its impact on their livelihoods, a small minority of the total number of long-haul truckers, accompanied by their supporters, descended on the nation's capital to vent their frustrations.[28] The protest was never very large: on the first day of the gathering in Ottawa, there were an estimated eight thousand protestors (compare this to a 2019 climate march in Montreal that involved five hundred thousand people).[29] By the following day, the numbers had dropped to fewer than half, with a continual decline over the life of the protest. The Freedom Convoy ended on February 23, after the imposition of the Emergencies Act, a seldom employed act that grants temporary power to the federal government to deal with emergencies it views as threats to the country's sovereignty

27. The title of this section comes from Richard Sanger's report on the Freedom Convoy, "Nothing Ever Happens in Ottawa," *London Review of Books,* April 21, 2022.

28. The Canadian Trucking Alliance, a federation of provincial trucking associations, was quick to point out that "the vast majority of the Canadian trucking industry is vaccinated with the overall industry vaccination rate among truck drivers closely mirroring that of the general public." See "Canadian Trucking Alliance Statement to Those Engaged in Road/ Border Protest," https://cantruck.ca/canadian-trucking-alliance-statement -to-those-engaged-in-road-border-protests/.

29. Andy Riga, "500,000 in Montreal Climate March Led by Greta Thunberg," *Montreal Gazette,* September 28, 2019, https://www.cbc.ca /news/canada/montreal/get-a-unique-view-inside-and-above-montreal-s -half-million-climate-march-1.5301122.

and security. The convoy had a real impact, though owing less to its size or to the length of time it occupied Ottawa than to the X-ray it offered of the contemporary state of Canadian political life. The hesitation at all levels of government to intervene in this monthlong protest highlights the power of an emerging common sense that can transform all contestation into an affirmation of the reality of its paranoid outlook on the world.

There was always something off about the accepted narrative of the principal cause of the protest.[30] After all, it was the imposition of the U.S. vaccine mandate for cross-border travel that impeded nonvaccinated Canadian truckers from hauling goods across the border, not a law imposed by the Canadian federal government—a fact that convoy organizers conveniently avoided mentioning. The initial demands for an end to the government vaccine mandate very quickly transformed into a clarion call for prime minister Justin Trudeau to leave office. This was perhaps to be expected. Neither Tamara Lich nor B. J. Dichter—the organizers of the GoFundMe and GiveSendGo campaigns used to generate support for the convoy—was a trucker, nor was Ontario convoy organizer Jason LaFace, a member of the ultra-right group Canada Unity. All three had been associated with prior antifederal movements; Lich, for instance, was a member of the Maverick Party (previously known as Wexit), which advocated for the independence of Western Canada from the rest of the country. Read generously, the Freedom Convoy could be seen as an expression of general frustration against a government unable or unwilling to address the concerns of ordinary Canadians; when asked by media what the demands of the convoy were, Dichter said, "It's everything, everything."[31] But the reality is more ominous.

30. The most detailed counternarrative is offered by Tanner Mirrlees, "The Carbon Convey: The Climate Emergency Fueling the Far Right's Big Rigs," *Energy Humanities,* May 3, 2022, https://www.energyhumanities.ca/news/the-carbon-convoy-the-climate-emergency-fueling-the-far-rights-big-rigs.

31. Brianna Sacks, "GoFundMe Says the Viral Campaign for Canada's Trucker Protest Hasn't Violated Its Rule Even Though It Sure Seems Like It

In addition to a familiar Western separatist ideology (of the kind framed in relation to control over oil extraction), it's clear that the imperatives of the convoy were shaped, at least in part, by QAnon, conspiracy theorists, and a range of U.S.-based right-wing groups.[32] A hack of the donor list of the GiveSendGo campaign confirmed this: 55.7 percent of donors were from the United States, and they included prominent donors to Donald Trump's campaign—a far greater indication of U.S. involvement in Canadian politics than that found by the Public Inquiry into Anti-Alberta Energy Campaigns led by Allan (which concluded that foreign money given to NGOs opposed to oil sands extraction was insignificant).[33]

This convoy may have been framed in response to a vaccine mandate as unpopular as Bill Gates is in some parts of the country (according to Google Trends, these two topics share consonant levels of online searches by Canadians). But at its core, this protest was about fossil fuels. To mobilize its participants, the organizers depended on the persistence of a regional politics anchored in deep-seated political disputes originating from the NEP. The right-wing ability to generate legitimacy for its positions by sliding from federal to provincial and back again allowed it to maintain a persistent confusion about the coronavirus pandemic. The federal government was held wholly culpable for an issue that fell under provincial jurisdiction: under the terms of the Canadian Constitution, provinces are responsible for the provision of health. In the wake of its quick abandonment of the plight of the unvaccinated truckers, one of the main demands organizers of the convoy made was for the governor general to dissolve the government, which she cannot do unless the government has lost the confidence of the House of

---

Has," *BuzzFeed,* February 3, 2022, https://www.buzzfeednews.com/article /briannasacks/gofundme-canada-vaccine-trucker-convoy.

32. Justin Ling, "5G and QAnon: How Conspiracy Theorists Steered Canada's Anti-vaccine Protests," *Guardian,* February 8, 2022, https://www .theguardian.com/world/2022/feb/08/canada-ottawa-trucker-protest -extremist-qanon-neo-nazi.

33. See note 16.

Commons.[34] It's unlikely that anything would have been gained by suggesting to Lich, Dichter, and convoy participants that they read up on the constitution. No argument could possibly cut the ideologically taut Gordian knot that binds two contradictory ideas together: the rejection of any and every action of government ("it's everything, everything") *and* the demand that the selfsame government act to undo itself; the hope that government might be able to undo the wrong it represents due to its very existence is further complicated by the starting premise that it can only ever manage to fuck things up.

The Canadian politics of fossil fuel can be seen as key to the convoy in even more direct ways. The 2022 Freedom Convoy built on the lessons of an earlier truck convoy to Ottawa. In February 2019, United We Roll brought hundreds of truckers to Ottawa to protest the federal government's introduction of a carbon tax and the development of strengthened environmental assessments of energy projects, as well as its ban on oil tankers on the northern coast of British Columbia. Some of those at the 2019 protest were also there to express their displeasure at the drop in oil prices (and drop in profits and job numbers)—a bizarre demand, if not for the view held by many Canadians that the government has at least some control over prices.[35] United We Roll was the concluding act of Canada's own version of the *gilets jaunes* protests, which saw yellow-vested Canadians protesting simultaneously against greenhouse gas policies and cultural threats posed by Muslims, immigrants, and the LGBTQ community.[36] More recently, there have

34. It is worth noting that the original memorandum issued to the governor general to dissolve the government contained no mention of truckers.

35. The West Texas Intermediate cost per barrel of oil in February 2019 was US$49.27. Oil from the oil sands is priced at a discount to WTI prices. See https://www.eia.gov/dnav/pet/hist/LeafHandler.ashx?n=PET&s =F003048623&f=M.

36. For an account of the French *gilets jaunes,* see Stathis Kouvelakis, "The French Insurgency," *New Left Review* 116/117 (March/June 2021): 75–98. Lich was one of the organizers of the yellow vests in Canada.

been calls for yet another convoy, this one to protest the Trudeau government's announced aim of a 30 percent absolute reduction in nitrous oxide emissions from fertilizer use. This has been refigured by some alt-right media outlets as an initiative to ban the use of *any* fertilizers. According to Joseph Quesnel, a research associate with the right-wing Frontier Centre for Public Policy, a new convoy—the third in three years—"could be fatal for our climate-obsessed PM."[37] Finally, the importance of oil for life in Canada is affirmed in the very form taken by the Freedom Convoy—a protest built out of trucks and saturated by the petro-masculinity and petro-populism connected with convoys ever since the release of the 1978 film of the same name.[38]

The confusing landscape of popular politics I've just painted isn't unique to Canada, even if the degree to which the right has connected it with the oil industry might be. In his 1995 essay "Ur-Fascism," Umberto Eco argues that fascism is grounded in a "cult of tradition," but a tradition that is syncretistic, a combination of different forms of belief and practice that "must tolerate contra-diction."[39] In Canada, fossil fuels have shaped right-wing notions of tradition that have proven able to easily navigate such contra-dictions. The implications of the right's political syncretism are obvious: nothing needs to be done with the natural world as long it remains the place where money magically flows from the ground;

37. Joseph Quesnel, "The Next Convoy Could Pull the Climate Curtains," Frontier Centre for Public Policy, September 3, 2022, https://fcpp.org/2022/09/03/the-next-convoy-could-pull-the-climate-curtains-on-trudeau/.

38. See Cara Daggett, "Petro-masculinity: Fossil Fuels and Authoritarian Desire," *Millennium* 47, no. 1 (2018): 25–44, and Caleb Wellum, "'Keep Moving': *Convoy* (1978), Car Films, and Petro-populism in the 1970s," in *American Energy Cinema,* ed. Robert Lifset, Raechel Lutz, and Sarah Stanford-McIntyre, 257–71 (Morganstown: University of West Virginia Press, 2023).

39. Umberto Eco, "Ur-Fascism," *New York Review of Books,* June 22, 1995, https://www.nybooks.com/articles/1995/06/22/ur-fascism.

to do anything about climate change is to be both anti-Canadian and dangerously opposed to tradition.

From the House of Commons to the Commons

This is the moment in a lecture series, in the denouement, when the speaker is expected to offer solutions to the various problems identified. I find myself hesitant to do this. I hope my analysis of these emergent narratives of energy transition, each of which makes a case that a specific variant of an energo-political status quo should be the one to determine what comes next, offers ideas for critical pathways to follow and other phenomena to assess and analyze. But I also recognize that the genre of a lecture series demands, at a minimum, an identification of some takeaways and next steps. I offer these tentatively and with some caution, not only because of the brief time remaining, but also because I am aware of how easy it can be to misread still emergent forces and positions.

To the repeated affirmation of the capacities of nations and nationalisms to address climate change by working together as a global collective of good liberal states, we should insist that a just energy transition will happen only through the creation of new international structures and systems truly committed to addressing climate change. Unlike existing international political mechanisms or organizations, such as the UN Environment Programme, these new structures would have the capacity to make decisions—even messy, inexact, or sometimes incorrect ones (because errors are inevitable and important for eventual successes)—on behalf of the planet. No such international structure exists at present in relation to energy or the environment, and one doesn't seem to be waiting in the wings. In *The Ministry for the Future* (one of the "hot reads" Bill Gates identifies on his website and the only one he doesn't entirely like), Kim Stanley Robinson may have come closest to imagining what such a structure could look like, in his account of how the world's central banks, each employing its institutional independence from the state it represents, manage to create an economic system in which the environment is no longer an unaccounted-for externality and markets reward government, business, and industry for climate action.[40] I recognize that this vision of effective international climate actors—national banks lead the revolution!—is far afield from those typically imagined within left environmental politics, but it has the virtue of challenging us to consider the potential of ideologically messy pathways to arrive at the internationalism that the climate crisis demands.

To meta-entrepreneurs and other powerful actors independent of state oversight and control, we need to articulate the desirability and necessity of a commons. Numerous thinkers have recently offered new accounts of the commons (or common in the singular); the best of them have tried to provide models for exactly what mechanisms might enable such new forms of political relationality to come into existence. Chantal Mouffe, for example, has argued that a left populism, fueled by the kind of strong group commitments that underlie contemporary right populism, could be constituted "around a *patriotic identification* with the best and more egalitarian aspects of the *national tradition.*"[41] This sounds eerily like the way the good liberal nationalisms I critiqued in my first lecture intend to mobilize their citizens to take on the extranational challenge of climate change. Pierre Dardot and Christian Laval have described the common as "in the most systematic and profound manner possible, the widespread introduction of institutional *self-government.*"[42] This is what many of us want and hope for when we imagine a new

40. Kim Stanley Robinson, *The Ministry for the Future* (London: Orbit Books, 2020).

41. Mouffe, *For a Left Populism,* 71, emphasis original. For the dangers and limits of developing a left politics through the use of populism, see Kai Bosworth, *Pipeline Populism: Grassroots Environmentalism in the Twenty-First Century* (Minneapolis: University of Minnesota Press, 2022), and Geoff Mann, "Who's Afraid of Democracy?," *Capitalism Nature Socialism* 24, no. 1 (2013): 42–48.

42. Pierre Dardot and Christian Laval, *Common: On Revolution in the Twenty-First Century,* trans. Matthew Maclellan (London: Bloomsbury, 2019), 314, emphasis original.

common. But although it's a laudable goal, it strikes me as very hard to reach in the political circumstances in which we currently find ourselves, and even dangerous if the ideal of self-government needs to emerge via patriotism. The world is replete with examples of patriotism foregrounding violent, exclusionary, and military nationalisms; there are far fewer examples of patriotism leading to the kind of internationalism necessitated by climate change. (World War II, the example of such a transmutation offered by some environmentalists, only holds up if one refuses to look at the deeper consequences of its patriotic fervor.)

As a rejoinder to the climate ideologies developed by meta-entrepreneurs, there may be a way forward that sidesteps some of the challenges inherent in the models and strategies articulated by Mouffe and Dardot and Laval. This is a notion of the common that foregrounds choices we have made to treat natural resources as property. George Caffentzis has pointed out that water and oil are both natural resources important to life on the planet.[43] Attempts to turn water into property have been (in the main) successfully resisted, as it seems obvious that no one should possess ownership over a resource so essential to life. Shouldn't oil and other resources—including the energy of the sun and wind—be challenged on these same grounds? The energy transition now under way is shaking up the old rules of resource ownership, which is precisely why there is a concerted struggle to lay claim to the common sense about the future of the sun on the part of those who want different variants of the energy status quo to prevail. Resource commoning in a post-oil world would put pressure on the continued legitimacy of the "tragedy of the commons," a narrative that continues to be used to assert the historical necessity of property and capitalism.[44]

43. George Caffentzis, "The Petroleum Commons," *Counterpunch,* December 15, 2004, http://www.counterpunch.org/2004/12/15/the -petroleum-commons/.

44. See Ian Angus, "The Myth of the Tragedy of the Commons," *Climate and Capital,* August 25, 2008, https://climateandcapitalism.com /2008/08/25/debunking-the-tragedy-of-the-commons/.

An emphasis on who will own energy resources and why an infinite source of energy needs to be owned at all may be a concrete way of laying the groundwork for the expansive political possibilities envisioned by Mouffe, Dardot and Laval, and others intent on bringing the exclusions and violence of extractive capitalism to an end. Who owns the sun? Everyone! Anyone who says otherwise must be challenged in the strongest terms possible.

This leaves me to account for the ideologues whose positions I have probed in this final lecture. What might be an appropriate rejoinder to their ideologies of resource extraction—an answer, at once conceptual and political, that would effectively unnerve the attempt of the right to lay claim to common sense on energy and our climate futures? The obvious answer might seem to be science, a discourse imagined to be able to puncture through the fictions and conspiracies on which the right depends to strengthen and grow. But science has proven to be limited in its capacities to articulate a new common sense, in part because it positions itself entirely outside of the political spectrum (neither right, left, nor center), and in part because it can so easily be framed as a fiction articulated by government or industry to suppress rights and freedoms; Gates is figured as a representative of science, as is Dr. Anthony Fauci and other senior health officials tasked with developing programs to manage the coronavirus pandemic. Science has, unfortunately, *animated* right-wing conspiracies rather than bringing unbelievers into the fold; it has also, unfortunately, had a limited capacity to mobilize action on climate change, even among individuals and polities that recognize the import and significance of its findings.

On what grounds does Mouffe believe we should turn to the "more egalitarian aspects of the *national tradition*," to create new forms of belonging that would unnerve everything the right tries to negate: difference, justice, and equality? I've suggested that her appeal to the nation and to patriotism is problematic, as there is little to guarantee that the egalitarian dimensions of nationalism can be put to different, better ends. But what about *tradition*—the category that lies at the heart of right-wing populisms and fascisms, the lost object that reactionaries wish to recover? Environmental discourse inevitably stresses the new, that is, what must come-into-being to adequately address the threat of climate change generated by extant ways of doing things. Part of the failure to date of the environmental movement to fully achieve its goals as quickly as the severity of the climate crisis demands lies in the threat posed by rapid, destabilizing change. As a result, however compelling degrowth (for example) might be as a political philosophy, it has a long and challenging path to travel before it might ever become common sense. The indeterminacy and unpredictability of the new, imagined by some as emerging from government conference rooms and via back-of-the-envelope plots by meta-entrepreneurs, can't help but falter against the comfortable weight of the quotidian, however damaged, desperate, and unhappy a place it might be; this is the lesson of Lauren Berlant's "cruel optimism."[45]

Tradition can be exclusionary, violent, degrading, and demoralizing. One might thus be forgiven for thinking that history has little in it to mount substantial challenges to the narratives I've examined here. But remember, tradition is not the same as history. It can be shaped to many ends, as the right knows so well. Tradition makes a demand on the present that can be harder to gainsay than threats posed by an unknown future. Consider the kinds of ideas that might be folded into tradition to challenge the legitimacy of status quo narratives: the common comes before capitalism; community precedes divisions and separations; the nation is little more than a savage fiction; and air, water, sky, and earth—and certainly the energy of the sun!—belong to everyone. Raymond Williams points to tradition as the most powerful mechanism of social incorporation and as fundamental to all politics. For Williams, it is a way in which "from a whole possible area of past and present, in a particular culture, certain meanings and practices are selected for

45. Lauren Berlant, *Cruel Optimism* (Durham, N.C.: Duke University Press, 2011).

emphasis and certain other meanings and practices are neglected or excluded."[46] The problem is thus not tradition per se but seeing it as fixed and unshakable—something about which there is nothing more that can be said, because the passage of time has made all that has come to be into a common sense that cannot be challenged. Conceding tradition to the status quo means forgoing its potential as a powerful political tool to unnerve the powers that be.

The environmental movement, and the left more generally, has tended to see much of what constitutes tradition today as little more than the "surviving past," filled with disappointments and dangers.[47] It would be wrong, of course, to suggest there is no tradition of left political achievement, of gains made and goals reached, that acts as a resource for contemporary political action and activism. Given what much of the surviving past looks like, progressives can be forgiven for avoiding it, perhaps especially those wishing to insist on the necessity of quick action on energy transition and climate change: the past is a record of climate failure, and what's needed now is climate success. But along with Williams, I think it is essential for those committed to climate action to make a claim on tradition, even if only as political rhetoric—a means of mythmaking to political ends. I have in mind here a deployment of tradition that might address Roland Barthes's criticisms of the limits of contemporary left-wing politics, which he sees as emerging from its reluctance to articulate new social myths, even as it labors to take apart the myths of others. Barthes writes,

> Left-wing myth is inessential: the objects which it takes hold of are rare—only a few political notions—unless it has itself recourse to the whole repertoire of the bourgeois myths. Left-wing myth never reaches the immense field of human relationships, the very vast surface of "insignificant" ideology. Everyday life is inaccessible to it: in bourgeois society, there are no "Left-wing" myths concerning marriage, cooking,

46. Raymond Williams, *Marxism and Literature* (1977; repr., Oxford: Oxford University Press, 1995), 115–16.

47. Williams, 115.

the home, the theatre, the law, morality, etc. Then, it is an incidental
myth, its use is not part of a strategy, as is the case with bourgeois
myth, but only of a tactics, or, at worst, of a deviation; if it occurs, it is
as a myth suited to a convenience, not to a necessity.[48]

Barthes characterizes the left as, at best, a kind of anti-right, dependent on the ideas of the right even as it rejects them. As such, left
politics can at times amount to little more than a softened myth of
the right—Mouffe's appeal to the national egalitarianism offers one
example of this, liberalism another. What may be missing on the left
is the *active* use of myth and the employment of its capacity to shape
tradition as strategy or tactic, in ways careful not to dismiss the present as nothing other than evidence of historical failure—a challenging political balancing act, to be sure. *Mythologies* was published
in 1957, at a time when what constituted left politics was very different than it is today (i.e., before the civil rights movement, global
challenges to colonial power, contemporary feminisms, Indigenous
struggles, Black Lives Matter, and the twenty-first-century environmental movement). Even so, his criticism of the left's capacities for
critique and discomfort with myth and tradition still has some bite.
To be clear, I am not suggesting that the use of myth and tradition
by groups and individuals engaged in climate actions is completely
absent. My comments are intended to draw attention to a tendency
I see that has made it difficult to sharply challenge the narratives
of energy transitions I've described here, which are shameless and
aggressive in their reconstitution of tradition to their own ends.

Just how might this use of tradition work? Before the Freedom
Convoy came the United We Roll convoy; they were animated by
similar aims and intentions, and the former drew considerable political energy from the example of the latter. But another convoy
headed toward Ottawa well before the right-wingers guzzled gas on
their way to Parliament Hill. In June 1935, one thousand residents

48. Roland Barthes, *Mythologies,* trans. Annette Lavers (New York:
Noonday Press, 1972), 147.

of a federal unemployment relief camp left Vancouver to take their protest over poor living conditions to Ottawa. Those in the On-to-Ottawa Trek commandeered freight trains to speed their process across the continent. The train convoy picked up strikers along the way, doubling its size by the time it reached Regina, where further progress was blocked by order of the prime minister, R. B. Bennett. Talks to address the strikers' concerns were initiated with the office of the prime minister but quickly broke down. Bennett's decision to arrest the trekkers resulted in the Regina Riot, which brought the trek to an end. It also brought an end to Bennett's career and fueled interest in the Canadian Communist Party, which had provided support to the trek.

Might one critique of the Freedom Convoy be that it dishonored the tradition of the On-to-Ottawa Trek, which was made up of men wanting to be paid more than the Can$0.20 per day they received for their work—a far cry from the convoy's inchoate demands for the prime minister to resign, just because a few people wanted it so? Might this have offered those who participated in the recent convoy, many of whom have been left behind politically and economically by neoliberalism, new insight into the cognate injustices that have been visited upon them today? And might there be a way of seeing Canada's NEP in the 1980s not as a failure but as a program intent on honoring a commitment to equality among Canadians, including equality with respect to their resources? In the era of climate change, the reinstitution of a program like the NEP (which, to be sure, was not without its problems) could change the direction of resource politics in Canada; instead, the federal government has chosen to support industry by, for example, purchasing the controversial Trans-Mountain pipeline from Kinder Morgan in 2018 for Can$4.5 billion. A new program to bring about energy transition—and fast—would bring us nowhere close to the common imagined in political philosophy. But it is a step in a process that may well be crucial to achieving real energy transition and truly addressing the causes and consequences of climate change.

The power of common sense comes from the immediate, intuitive claim it can make to speak on behalf of the center—that apparently nonideological, inclusive, and democratic space in which the only statements made are ones on which everyone agrees. In making this claim, common sense asserts its right to speak for the most salable aspects of tradition. The end of the oil era does not denote the end of capitalism. Nor does the expanding use of renewable energy produce (ipso facto) new, far better ways of living together and in the world: a world of clean energy can fuel a world still shaped by racism, sexism, and discrimination based on class, gender, ethnicity, sexual orientation, disability, and more. The future of the sun won't depend on the technological sophistication of solar panels and the storage capacity of batteries. It will depend on our willingness to forcefully challenge those narratives of energy transition that promise change while insisting that we keep everything exactly as it is. It will depend, too, on our ability to tell convincing stories about the traditions we need and want to honor—and those to which it is no longer worth paying any attention, because they have gotten us nowhere.

Acknowledgments

These lectures owe their existence to Rhys Williams, who invited me to spend three months working on them at the University of Glasgow in summer 2022, and the Leverhulme Trust, which awarded me the fellowship that made that trip possible. In Glasgow, my friend and colleague Graeme Macdonald was an invaluable interlocutor. I miss our daily midafternoon coffee sessions, where we talked energy ideas and political ideals. Mark Simpson's editorial suggestions were, as usual, superb. I owe everything good in these pages to Eva-Lynn Jagoe, Tanner Mirrlees, and Andrew Pendakis—all of them always know the right thing to say. And an enormous thanks to Valerie Uher for helping on this project, just as she has on so many others over the past six years.

(Continued from page iii)

Forerunners: Ideas First

Alexis L. Boylan, Anna Mae Duane, Michael Gill, and Barbara Gurr
Furious Feminisms: Alternate Routes on *Mad Max: Fury Road*

Ian G. R. Shaw and Marv Waterstone
Wageless Life: A Manifesto for a Future beyond Capitalism

Claudia Milian
LatinX

Aaron Jaffe
Spoiler Alert: A Critical Guide

Don Ihde
Medical Technics

Jonathan Beecher Field
Town Hall Meetings and the Death of Deliberation

Jennifer Gabrys
How to Do Things with Sensors

Naa Oyo A. Kwate
Burgers in Blackface: Anti-Black Restaurants Then and Now

Arne De Boever
Against Aesthetic Exceptionalism

Steve Mentz
Break Up the Anthropocene

John Protevi
Edges of the State

Matthew J. Wolf-Meyer
Theory for the World to Come: Speculative Fiction and Apocalyptic Anthropology

Nicholas Tampio
Learning versus the Common Core

Kathryn Yusoff
A Billion Black Anthropocenes or None

Kenneth J. Saltman
The Swindle of Innovative Educational Finance

Ginger Nolan
The Neocolonialism of the Global Village

Joanna Zylinska
The End of Man: A Feminist Counterapocalypse

Robert Rosenberger
Callous Objects: Designs against the Homeless

William E. Connolly
Aspirational Fascism: The Struggle for Multifaceted Democracy under Trumpism

Chuck Rybak
UW Struggle: When a State Attacks Its University

Clare Birchall
Shareveillance: The Dangers of Openly Sharing and Covertly Collecting Data

la paperson
A Third University Is Possible

Kelly Oliver
Carceral Humanitarianism: Logics of Refugee Detention

P. David Marshall
The Celebrity Persona Pandemic

Davide Panagia
Ten Theses for an Aesthetics of Politics

David Golumbia
The Politics of Bitcoin: Software as Right-Wing Extremism

Sohail Daulatzai
Fifty Years of *The Battle of Algiers*: Past as Prologue

Gary Hall
The Uberfication of the University

Mark Jarzombek
Digital Stockholm Syndrome in the Post-ontological Age

N. Adriana Knouf
How Noise Matters to Finance

Andrew Culp
Dark Deleuze

Akira Mizuta Lippit
Cinema without Reflection: Jacques Derrida's Echopoiesis and Narcissism Adrift

Sharon Sliwinski
Mandela's Dark Years: A Political Theory of Dreaming

Grant Farred
Martin Heidegger Saved My Life

Ian Bogost
The Geek's Chihuahua: Living with Apple

Shannon Mattern
Deep Mapping the Media City

Steven Shaviro
No Speed Limit: Three Essays on Accelerationism

Jussi Parikka
The Anthrobscene

Reinhold Martin
Mediators: Aesthetics, Politics, and the City

John Hartigan Jr.
Aesop's Anthropology: A Multispecies Approach

Imre Szeman is director of the Institute for Environment, Conservation, and Sustainability and professor of human geography at the University of Toronto Scarborough. He is the cofounder of the Petrocultures Research Group and the founder of the After Oil Collective.